Refining the committee approach and uncertainty prediction in hydrological modelling

Refining the committee approach and uncertainty prediction in hydrological modelling

DISSERTATION

Submitted in fulfilment of the requirements of
the Board for Doctorates of Delft University of Technology
and of the Academic Board of UNESCO-IHE
Institute for Water Education

for
the Degree of DOCTOR
to be defended in public on
Thursday, 30 October 2014, at 15:00 hours
in Delft, the Netherlands

by

NAGENDRA KAYASTHA

Master of Science Water Science and Engineering
specialization in Hydroinformatics
UNESCO-IHE Institute for Water Education, Delft, the Netherlands

born in Bhaktapur, Nepal

This dissertation has been approved by the promotor:
Prof. dr. D. P. Solomatine

Composition of Doctoral Committee:

Chairman	Rector Magnificus Delft University of Technology
Vice-Chairman	Rector UNESCO-IHE
Prof. dr. D. P. Solomatine	UNESCO-IHE / Delft University of Technology, promotor
Prof. dr. ir. A. B. K. van Griensven	Vrije Universiteit Brussel / UNESCO-IHE
Prof. dr. ir. W. G. M. Bastiaanssen	Delft University of Technology / UNESCO-IHE
Prof. dr. ir. A. E. Mynett	UNESCO-IHE / Delft University of Technology
Prof. dr. G. Di Baldassarre	Uppsala University, Sweden
Prof. dr. ir. P. Willems	University of Leuven, Belgium
Prof. dr. ir. M. Kok	Delft University of Technology (reserve)

CRC Press/Balkema is an imprint of the Taylor & Francis Group, an informa business

Published by:
CRC Press / Balkema
PO Box 11320, 2301 EH Leiden, The Netherlands
e-mail: Pub.NL@taylorandfrancis.com
www.crcpress.com - www.taylorandfrancis.com - www.ba.balkema.nl

ISBN 978-1-138-02746-6 (Taylor & Francis Group)

To my mother, wife and sons
To the memory of my beloved late father

SUMMARY

A hydrological model is an abstraction of complex and non-linear physical processes that operates to predict the behaviour of time-varying streamflows in a catchment. The strength of such predictions depends on the presumed model structure, the described parameters, and the quality of data used. Generally, model predictions assume that data fed into hydrological model (conceptual lumped) and its overall structure are correct, and model prediction is deliberately presented based on measurement data using degree of knowledge by discovering the optimum parameter set (calibration). However, the model predictions need to consider subsequent uncertainty because calibration and uncertainty procedures are associated with each other. The confidence of model outputs cannot be dealt without evaluation of uncertainty that represent a prediction of hydrological responses.

Often single hydrological models cannot equally describe the characteristics of hydrological processes for all ranges of model outputs (streamflows), due to the multiple hydrological responses and their value in different magnitudes. The multi modelling approach opens up possibilities for handling such difficulties and allows improve the predictive capability of models. One of multi modelling approaches called "committee modelling" is one of the topics in part of this study. In this approach, different individual models specialized on distinctive hydrological regimes that instantiated in same model structure are combined to produce a single new model where each individual model's strength is presented optimally and their weaknesses compensated by each other.

Special attention is given to the the so-called "fuzzy committee" approach to hydrological modelling (Solomatine, 2006; Fenicia et al. (2007). In it first different processes (range of catchment responses) are calibrate which fit to represent a particular process and merge them through a fuzzy weighing. Further tests using this approach have been carried out by Kayastha et al. (2013) with proposing several types of weighting schemes in objective functions to calibrate the specialized models, as well as different classes of membership functions to combine these models. The models are built for different components of flow hydrograph separately and then combined using appropriate methods to provide a more comprehensive and accurate prediction. Such models referred to "committee models" in this thesis. The weights assigned to each specialized model's output are based on optimally designed fuzzy membership functions. The results of experiments are presented in this thesis. All the committee models have shown a good efficiency in model predictions compared to single hydrological (optimal) models, which are applied for prediction of conceptual hydrological model for the Alzette, Bagmati, Brue, and Leaf catchments. In addition, the test results of these newly proposed committees models where weights are calculated based on model state variables (soil moisture, base flow, etc.), inputs (precipitation and evapotranspiration) and outputs (simulated streamflows) are also reported here and these weights are different at every time depending on the current value of flow.

The models specialized on low flows of the catchments have a relatively high error compared to high flows. One possible way to improve the performances of the overall committee model is using hybrid committee models. In this approach, a committee model is formed from two specialized models (conceptual model for high flows and the data-driven artificial neural networks model for low flows) using an appropriate combination method (fuzzy membership function). Hybrid committee models are tested in the Bagmati and Leaf catchments and their it has been found that they are the most accurate among all committee models.

Another important theme addressed in this study is uncertainty analysis and prediction. Uncertainty analysis helps to enhance the reliability and credibility of model predictions in hydrological modelling. One aspect here relates to Monte Carlo (MC) simulation widely used for uncertainty analysis. In it the model outputs associated with a set of inputs or/and parameters obtained from the given distributions and then a quantitative estimate of the confidence is computed. Generally, this needs a large number of model simulations and therefore more attention has to be given to developing the economical sampling schemes that allow working with computationally intensive models. This thesis presents the results of the investigated effects of different sampling schemes (MCS, GLUE, MCMC, SCEMUA, DREAM, PSO, and ACCO) for uncertainty estimations of hydrological models. Comparative interpretation of the resulting uncertainty statistics shows that the uncertainty analysis using sampling in Monte Carlo framework should take into account that the uncertainty estimates considerably depend on the sampling method used.

Another aspect of uncertainty analysis relates to predicting uncertainty (rather than its analysis). The estimation of uncertainty based on MC simulations methods is generally valid for the past data, however, it is not necessarily valid for the future model runs in operation. To overcome such difficulties, it would be beneficial to find economical ways of predicting uncertainty for the future states of an environmental system. Machine learning techniques (data-driven modelling) are used to improve the accuracy of hydrological model prediction/forecasting, however, these techniques do not permit to build probability distribution function of model (model uncertainty). Shrestha et al. (2009, 2013) proposed to build model of probability distribution function as predictive uncertainty models, which allows an adequate uncertainty estimation for hydrological models. Inputs to these models are specially identified representative variables (past events precipitation and flows, and possibly soil moisture). The trained machine learning models are then employed to predict the model output uncertainty, which is specific for the new input data. A brief description of a method to access uncertainty of the model by encapsulating and predicting parameter uncertainty of hydrological models using machine learning techniques and their results are presented in this thesis. This method is tested in the Bagmati and Brue catchments, to predict uncertainty (quantiles of *pdf*) of the deterministic outputs from the HBV conceptual hydrological model. The results reveal that this method is efficient for assessing uncertainty and produced results are quite accurate. Furthermore, this method is tested with various sampling based methods of uncertainty outputs of hydrological models.

The results produced by several predictive uncertainty models (machine learning models) vary, the reasons being: a) the use of different predictive uncertainty models that is results of various sampling algorithms in different data sets used to train a predictive uncertainty model, and b) different sets of inputs data used to train a predictive uncertainty model which leads to several models. In this thesis a combination of models is proposed (forming thus a committee) which is applied to estimate the uncertainty of streamflows simulation from a conceptual hydrological model in the Bagmati and Nzoia catchments.

An important "user" of hydrological models' simulations is flood inundation modelling. The associated uncertainty provides additional information for decision making that is related to preparedness and minimizing losses from flooding. This modelling process requires information on river flows (e.g, boundary conditions, Manning's coefficients, channel cross section and depth), observations of flood extent (topographic data), and method for quantifying the performance of the flood inundation pattern. Runoff is the main contributor to flood hence knowledge on flow characteristics of a certain flood event also required for modelling of inundation. The complexity of flood processes can be represented by forming a sequence (cascade) of models (hydrological and hydraulic) and by geospatial processing. However, such integration is not easy to set up, because it requires large amounts of data, processing power and knowledge on the process interactions between models. Various sources of uncertainty have to be considered which resulting in uncertain model cascade outcomes. One common method to estimate uncertainty is MC technique, which is used to produce an ensemble of deterministic model simulations and then assigning it the goodness of fit measure based on observed flood inundation extent. Remotely sensed data (maps) of flood extent can be to calibrate the models in a deterministic framework with a single observed event.

A realistic uncertainty analysis of such integrated models requires multiple model runs and hence enough computational resources. In this thesis, SWAT hydrological and SOBEK hydrodynamic models are integrated (cascade) to quantify of the uncertainty in flood inundation extent for the Nzoia catchment in Kenya. These models are set in the high performance computing framework (parallel computing on a cluster) and the final outputs used to estimate the uncertainty in flood inundation extent which is presented as the relative confidence measure.

Overall, this thesis presents research efforts in: (i) committee modelling of hydrological models, (ii) hybrid committee hydrological models, (iii) influence of sampling strategies on prediction uncertainty of hydrological models, (iv) uncertainty prediction using machine learning techniques, (v) committee of predictive uncertainty models and (vi) uncertainty analysis of a flood inundation model. This study is a contribution to hydroinformatics, which aims to connect various scientific disciplines: hydrological modelling, hydrodynamic modelling, multi-model averaging, data driven models, hybrid hydrological models, uncertainty analysis and high performance computing. The drawn conclusions allow for advancing the theory and practice of hydrological and integrated modelling. The developed software is made available for public use and can be used by the researchers and practitioners to advance the mentioned areas further.

Nagendra Kayastha

Delft, The Netherlands

Table of Contents

Chapter 1
Introduction

This chapter introduces existing research on committee modelling, intended to improve hydrological model prediction, uncertainty estimation of hydrological models, and uncertainty prediction using machine learning techniques. Committee models use a multi-modelling approach in an effort to improve model prediction by involving a combination of model outputs. An overview of committee of predictive uncertainty models, and uncertainty analysis of integrated models, is presented. Finally, research questions, objectives of research, a description of the catchments used in the present case studies, and an outline of this thesis are presented.

1.1   Background

Hydrological modelling tools are employed in a wide range of applications, for example, estimating flows of ungauged catchments, for real-time flood forecasting, in the design and operation of hydraulic structures, and to study the effects of land-use and climate change. Hydrological models attempt to describe rainfall-runoff relationships, and these relationships are very complex due to non-linear and spatial-temporal variability of the rainfall process and catchment characteristics.

Despite the regular emergence of new models, and with a wide spectrum of existing models, no single model exists that demonstrates superior performance for all catchments (Nayak et al., 2005). The current trend is shifting towards building increasingly complex and sophisticated models because of rapid advancement in computational efficiency, as well as a better understanding of the physics and dynamics of water processes. Such complex and sophisticated models may still be inaccurate in representing reality, due to the use of multiple parameters and a lack of reasonable quality input data. Therefore, a model has to describe information by using the simplest useful structure possible, which would use reasonably accurate estimates of unknown model parameters and encompass good predictive capability.

A hydrological model is an abstraction of a complex, non-linear, time and space-varying hydrological process, attempting to imitate reality. This model operates to allow predictions of the behaviour of varying streamflows in the catchment over time. However, the validity of predictions always depends on the presumed model structure, parameters, and quality of data used. In usual practice, modellers often assumed that the data fed into hydrological models (conceptually lumped), and its overall structure are correct, and that the model prediction presents information based on the collected measurement data, using the identified optimal parameter set. However, there always exists an inconsistency between the model prediction and the corresponding measurement data, regardless of how precisely the model has been built and how perfectly the model is calibrated. The prediction of streamflows from hydrological models is persistently constrained by the following factors: (i) multiple hydrological responses, for instant, high flow, low flow and water balance (Kollat et al., 2012); (ii) one or more objectives to express the tradeoffs between the observed and simulated outputs (Zhang et al., 2009); and (iii) different performance measures are sensitive to different flow magnitudes (Westerberg et al., 2011). These constraints oblige modellers to improve model prediction by investigating a multi-modelling approach, which might involve multi-objective calibration (Efstratiadis and Koutsoyiannis, 2010), ensemble modelling (Viney et al., 2009), and model averaging (Ajami et al., 2006; Fenicia et al., 2007; Kayastha et al., 2013).

Looking at the history of hydrological modelling, advancements have made considerable changes to Sherman's unit hydrograph method (Sherman, 1932) towards conceptual models (e.g. HBV) and process-based models (e. g., MIKE-SHE). Moreover, data-driven (regression) models have been also successfully to describe the rainfall-runoff relationships.

Apart from dealing with model accuracy, hydrological modelling requires proper estimation and thoughtful interpretation of uncertainty in order to understand the significance of the results. Incorporating uncertainty into deterministic predictions or forecasts helps to enhance the reliability and credibility of the model. The realistic estimation of the corresponding predictive uncertainty helps in adequate decision-making processes (Georgakakos et al., 2004). There are three major sources of uncertainty in modelling: (i) errors in input data and data for calibration; (ii) deficiency in model structure; and (iii) uncertainty in model parameters. Monte Carlo (MC) techniques are commonly used to estimate the uncertainty of hydrological models, however these techniques use past data, so that the estimates are not necessarily valid for future model runs during operation. Hence, it would be beneficial to develop practical ways to estimate the model uncertainty for future situations.

In the context of flood management, hydrological models are typically linked to hydraulic modelling and geospatial processing, and these are carried out by integration of the hydrological and hydrodynamic (1D and 2D) models. However, such integration requires the accessibility of data, processing power, and complex process interactions between models. Models are always influenced by various sources of uncertainty, the study of uncertainty in flood modelling serves as important information for decision-making that relates to preparedness and for minimizing losses from flooding. Uncertainty analysis of integrated models based on Monte Carlo simulations requires considerable computational resources.

This thesis principally explores enhancements in committee hydrological models, hybrid hydrological models, various sampling strategies for uncertainty analysis, uncertainty prediction using machine learning techniques, committee of predictive uncertainty models, and flood inundation models and their estimation of uncertainty, using high performance computing. The corresponding literature review is provided in each chapter.

1.1.1 Conceptual hydrological models

Conceptual hydrological models are simplified representations of the hydrological processes in a catchment. These are composed of a number of fluxes and storages, and are described by mathematical equations. Storages are interconnected through fluxes of rainfall, infiltration, percolation, and other factors that control the way in which water is added, stored, transmitted, and discharged from the system, representing physical elements (White, 2003). The mathematical equations used to describe the system are semi-empirical, with a physical basis. Parameters and fluxes represent the average over the entire catchment. While these parameters cannot be measured in the field, they can be estimated through model calibration. However, for calibration to be accurate, there is a need for sufficient hydro-meteological records, which may not be continuously available. Well-structured conceptual models should be simple, easy to implement in the computer code, and reveal the model complexity and prediction capability.

Many conceptual hydrological models have been developed with different levels of physical representation and varying degree of complexity. However, to easily operate them, it is required that their characteristics be well understood. Crawford and Linsley (1966) introduced one of the first (widely cited) conceptual model called the Stanford Watershed Model. Other models include the TANK model (Sugawara, 1967, 1995); Sacramento Soil Moisture Accounting model (Burnash et al., 1973); NAM model (Nielsen and Hansen, 1973); HBV model (Bergström and Forsman, 1973); TOPMODEL (Beven and Kirkby, 1979); and others. A brief description of several early conceptual models can be found in Fleming (1975). In addition, Singh (1995) provided comprehensive descriptions of a large number of conceptual models.

1.1.2   Committee hydrological models (multi-models)

The complexity of most natural phenomena originates from the fact that they are composed of a number of interacting processes. However, their modelling is typically concentrated on a single model handling all processes without consideration for local solutions (Corzo and Solomatine 2007). Such simple, single-issue models have less prediction capability and often suffer from inaccuracies. The solution to this challenge could be an approach to modelling that handles different sub-processes separately with diverse models fit to represent a particular process. When the process-based modelling paradigm of modelling is used, every model can be built, specifically oriented to a particular process, or the same model structure can be used but calibrated differently for different regimes of the same process (Fenicia et al., 2007; Kayastha et al., 2013). In the case of data-driven models (for example, neural networks), the training set is split into a number of subsets, and separate models are trained on these subsets (Corzo and Solomatine 2007). The input (state) space can be divided into a number of regions in each of which a separate specialized model is built (Figure 1-1.). These specialized models are also called local or expert models, and form a modular model (MM). One of the issues to consider here is to ensure compatibility of the local models at the boundaries between the processes or regimes which can be done by using so-called fuzzy committees (Solomatine 2006).

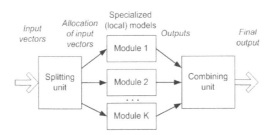

Figure 1-1. Modular modelling: local models are responsible for particular sub-processes and their outputs combined (Solomatine, 2006)

One of the challenges in conceptual hydrological models is to identify a set of parameters, characterizing the behaviour of time-varying streamflows in a catchment. In lumped models, the parameters cannot be measured directly, due to the problems related to dimensionality and scaling (Beven 2000). These are computed based on the measurement of meteorological forcing data to produce model predictions that are as close as possible to the observed discharge data, using some degree of expertise and experience. Typically, this approach focuses on the single model, using the single best set of parameters, and the committee approach assumes multiple models can be built from different components of streamflows that correspond to characteristics of different flow regimes. These models are then combined to provide a more comprehensive and accurate representation of catchment processes. Such models are referred to as multi-models, or committee models. The main hypothesis of the multi-model combination approach is that different models depict different aspects of the data. For this reason, the combination of these aspects can better predict streamflows than those produced by any one of the individual models involved in the combination (Fernando et al., 2012).

Recently, so-called model averaging has been receiving more attention in hydrological modelling. In essence, (weighted) model averaging is a subset of committee modelling focussing on particular ways of combining models. The idea of model averaging is also to integrate individual models into a single new model, where each individual model's strengths are presented in such way that one can obtain optimal prediction, with the weaknesses of each model compensated for by each other. This section of the thesis presents one of the ways to improve prediction of hydrological models by the modelling of different processes separately. Each model represents a particular process, and they can then be merged to produce a combined model having a higher degree of accuracy.

1.1.3   Uncertainty analysis of hydrological models

Several uncertainty analysis methods have been developed to predict the uncertainty of hydrological models and to derive meaningful uncertainty in model outputs. Broad classification of these methods can be found in Shrestha and Solomatine (2008). The Monte Carlo (MC) simulation technique has been used successfully for uncertainty analysis in hydrological modelling for many years. It allows for the quantification of model output uncertainty resulting from uncertain model parameters, input data or model structure. The approach involves random sampling from the distribution of uncertain inputs, and the model runs continuously until a desired statistically significant distribution of outputs is obtained. The main advantage of MC simulation is that it is simple to apply. However, it requires a large number of samples (or model runs), so their applicability may be limited only to simple (fast) models. In the case of computational intensive models, the time and resources required by this method could be prohibitively expensive.

One of the versions of MC analysis is the Generalized Likelihood Uncertainty Estimation (GLUE) [(Beven and Binley (1992), see also its critical analysis by Stedinger et al. (2008); and Mantovan and Todini (2006)]) - it is quite popular in hydrological studies. A procedure for partially correcting the prediction limit in the GLUE method has been proposed by Xiong and O'Connor (2008).

One of the canonical sampling methods is the Markov Chain Monte Carlo (MCMC) method (for hydrological applications, see Kuczera and Parent, 1998; Gilks et al., 1998; Yang et al., 2008). Vrugt et al., 2003 proposed merging the MCMC sampler with the SCE-UA global optimization algorithm, and Blasone et al. (2008a,b) proposed a version of GLUE based on MCMC sampling.

If a reliable method for MC simulation based uncertainty estimation of hydrological models is to be developed, one has to ensure that the results would not depend too much on the sampling method used. To facilitate meaningful interpretation of uncertainty results, it is necessary to investigate the effects of different sampling schemes for uncertainty estimations of hydrological models, to assess applicability of "economical" sampling schemes (allowing for working with computationally intensive models) and compare them.

1.1.4   Uncertainty analysis using machine learning techniques

Machine learning (ML) techniques (data-driven modelling) are widely used in the field of rainfall-runoff modelling to improve the accuracy of prediction/forecasting. They are also used to build emulators (surrogates) of the process-based models. Shrestha et al., 2009 proposed to use ML to build predictive models of uncertainty (this method is entitled Machine Learning for Uncertainty Estimation (MLUE)). These techniques do not permit the building of the probability distribution function of model output directly, but it is possible to build a model that would predict the quantiles of this function and thus allow for reasonable uncertainty predictions (Shrestha et al., 2009; Shrestha et al., 2013).

It has been already mentioned above about the advantanges of MC simulation-based uncertainty analysis techniques. These techniques provide only average measures of uncertainty based on past data. However, if one needs to estimate the uncertainty of a model in a particular hydro-meteorological situation in real-time application of complex models, MC simulation becomes impractical because of the large number of model runs required. In this respect, machine learning techniques can be used as predictive models that emulate the MC simulations and, hence, provide an approximate solution to the uncertainty analysis in a real-time application without re-running the MC simulations. This method allows for assessing uncertainty of complex models in real time. Part of this thesis explores an efficient method to assess the uncertainty of the model by encapsulating and predicting the parameter uncertainty of hydrological models, using machine learning techniques.

1.1.5   Committee of predictive uncertainty models

The MLUE method (Shrestha et al., 2009; Shrestha et al., 2013) allows for building the predictive uncertainty models that use the results of MC sampling (or any other sampling scheme) and are able to predict uncertainty (quantiles of *pdf*) of the deterministic outputs from hydrological model. The inputs to these models are specially identified representative variables (past events of precipitation and flows). The trained machine learning models are then employed to predict the model output uncertainty, which is specific for the new input data.

The problem here is that different sampling results in different data sets used to train a predictive uncertainty model, which results in several models. These numerous predictive uncertainty models (machine learning models) produce several uncertainty results in calibration and verification. However, the results presented from a group of competing models are much more complex than any single model. Each model has its own predictive capabilities and limitations. The combination of competing models allows the strength of each individual model to merge in an optimal way so that the best prediction can be obtained. Combining models require determining the weights, which average the model outputs thereby taking advantages of each individual model. Part of this thesis proposes a method that improves their prediction by merging their outputs optimally, that is, to form a committee of all predictive uncertainty models to generate the final output.

1.1.6 Flood inundation models and their uncertainty

Description of flood processes and their spatial representation is a complex and interdisciplinary task, which can be realized by understanding the hydrology and hydraulics of a system. Accordingly, integration of relevant models is necessary. The coupling of two or more models usually precedes such integation frameworks, where the outputs of one model provide inputs for another. This allows easy data transfer between models, —not only related to data but also for the associated uncertainties.

Even though it is not straightforward to apply, the associated uncertainty should be quantified for each model involved (Pappenberger and Beven, 2006; Todini and Mantovan, 2007; Beven, 2009). A common problem of such models is the challenge of quantifying and describing these uncertainties (McMillan and Brasington, 2008; Cloke and Pappenberger, 2009). Furthermore, there may be inconsistencies in the results because the individual models may describe the same processes in different ways, while most models have been designed to simulate specific aspects of the water processes (Guzha and Hardy, 2010). In the last few years, many efforts have been undertaken to deal with the integrated modelling in hydrological and hydrodynamic domains (He et al., 2009), but still their accuracy is an issue (Pappenberger et al., 2009). Despite the progress made in integrated modelling, links between the uncertainties of models have not been systematically explored, and the investigation of such a framework is important for uncertainty studies of integrated models.

When models are used for decision-making, it is, therefore, crucial that the uncertainties are properly described. In a linked modelling system, this is a real challenge due to the multiple sources of uncertainty. To gain insight into this problem, it is necessary to assess the uncertainties, when these have passed through the linked models. Availability of high-performance computers and cluster/cloud solutions makes it possible linking of complex models, and explicitly presenting the uncertainties associated with predictions.

1.2 Research questions

The key research questions addressed in this thesis are as follows:
- (a) How the committee modelling approach would allow for improving hydrological model prediction further?
- (b) How to improve methods of combining process-based and data-driven models (hybrid modelling), for the improvement of hydrological model accuracy?
- (c) How do various sampling strategies affect the uncertainty estimation of hydrological models?
- (d) How can machine learning models be tuned and applied for predicting hydrological model uncertainty?

(e) How is it possible to combine predictive uncertainty models?

(f) How can we propagate the uncertainties in the linked hydrological and hydrodynamic models in an efficient way?

1.3   Research objectives

The main objective of this research is to further develop methods to improve the committee modelling approach, and effective and efficient methods for uncertainty analysis of hydrological models. Specific objectives are:

- To refine the methodology of committee (multi-model) modelling, focussing on the dynamic weighted-averaging approach.
- To explore the possibilities of enhancing the accuracy of hydrological models by combining the processes-based and data driven models (hybrid models).
- To analyze the effects of different sampling strategies for estimation of uncertainty of a hydrological model.
- To further develop and refine the uncertainty analysis method MLUE (based on using machine learning to encapsulate the results of MC runs).
- To implement and test the multi-model averaging approach for predictive uncertainty models.
- To quantify and propagate uncertainty in a chain of hydrological and hydraulic models (on the Nzoia catchment case study).
- To implement the software integrating the SWAT modelling system with the NSGAX tool for multi-objective calibration.

1.4   Case studies

The descriptions of five different catchments taken for case studies in this research are given below.

1.4.1   Alzette catchment

Alzette catchment is located in the large part of the Grand-Duchy in Luxembourg. The river has a length of 73 km along France and Luxembourg. The streamflows are measured at Hesperange gauging station, which is placed along the Alzette River upstream of Luxembourg-city. The drainage area of the catchment is 288 km^2, and land cover is composed of cultivated land (27%) grassland (26%), forestland (29%) and urbanized land (18%). Marls and Marly-sandstones on the left bank tributaries and limestones on the right bank tributaries of the Alzette River mainly represent lithology. Marls areas are characterized by impermeable bedrock, therefore rainfall water, after losses for evaporation, reaches the stream mostly as saturated subsurface flow that develops at the interface between the weathered zone and the underlying bedrock areas. When the weathered zone becomes saturated, or during heavy rainfall events, surface runoff occurs. In limestone areas, a large part of rainfall water infiltrates and after subtraction of losses percolates to the groundwater aquifer, which is capable of storing and releasing large quantities of water. The response to rainfall of Marl areas is faster and characterized by larger volumes of water than that of limestone areas. Moreover, the large part of the baseflow during prolonged dry periods is mostly sustained by the limestone aquifer (Fenicia et al., 2006). The basin is instrumented by several rain gauges including tipping-buckets and automatic samplers measuring at a time step which does not exceed 20 min. Hourly rainfall series were calculated by averaging the series at the individual stations with the Thiessen polygon method. Daily potential evaporation was estimated through the Penman-Monteith equation (Monteith, 1965).

Figure 1-2. Location map of the Alzette catchment in Luxembourg, black triangles denote the rainfall stations, and circles denote the discharge gauging stations.

1.4.2 Bagmati catchment

Bagmati catchment (26°42′–27°50′N and 85°02′–85°58′E) lies in the central mountainous region of Nepal. The elevation ranges from 57 m to 2,913 m encompasses nearly 3700 km^2 within Nepal and reaches the Ganges River in India. The catchment area draining to the gauging station at Pandheradobhan is about 2900 km^2 (see Figure 1-3) and it covers the Kathmandu valley including eight districts of Nepal. The source of the Bagmati River is Shivapuri which is surrounded by Mahabharat mountain ranges at an altitude of around 2690 m. The length of the main channel is about 195 km within Nepal and 134 km above the gauging station. Discharge measured at Pandheradobhan is used for the analysis (adopted from Solomatine et al. (2008). The altitude discharge gauging stations elevations is 180 m and peak discharge is found to be 5030 m^3/s based on the data from 1988 to 1995. The mean daily discharge is approximately 150 m^3/s measured with average precipitation of 250 mm and air temperature is 15.8 ∘ C. More than half of the watershed area (58%) is covered by forest. Cultivated land accounts for 38% of the area of the watershed while nearly 4% of the land in the watershed is barren. Most of the area of this catchment is occupied by the hilly and mountainous land. The mean areal rainfall was calculated using Thiessen polygons. Although this method is not recommended for mountainous regions, the mean rainfall is consistent with the long-term average annual rainfall which is computed by the isohyetal method (Chalise et al., 1996). The long-term mean annual rainfall of the catchment is about 1500 mm with 90% of the rainfall occurring during the four months of the monsoon season (June to September). Hydrological seasons are categorized in three different groups in Nepal: (a) dry pre-monsoon season (March–May) with almost no rain; (b) rainy monsoon season (June–September) and (c) post-monsoon season (October–February) with little rain.(Sharma and Shakya, 2006).

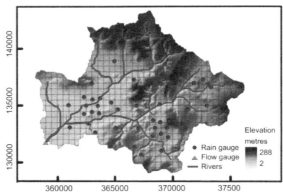

Figure 1-3. Location map of the Bagmati catchment in Nepal, triangles denote the rainfall stations, and circles denote the discharge gauging stations.

1.4.3 Brue catchment

The Brue catchment is located in South West of England, UK. This catchment has been extensively used for research on weather radar, quantitative precipitation forecasting and rainfall-runoff modelling, as it has been facilitated by a dense rain gauge network as well as coverage by three weather radars. Numerous studies (Bell and Moore, 2000; Moore, 2002) have been conducted regarding the catchment, especially by the Hydrological Radar EXperiment (HYREX),Special Topic Program of Natural Environment Research Council (NERC), UK. Figure 1-4 shows the locations of the Brue catchment and the gauging stations. The major land use is pasture on clay soil and there are some patches of woodland in the higher eastern part of the catchment.

Figure 1-4. The Brue catchment showing dense rain gauges network (the horizontal and vertical axes refer to the easting and northing in British national grid reference coordinates).

The catchment has a drainage area of 135 km^2 with the average annual rainfall of 867 mm and the average river flow of 1.92 m^3/s, for the period from 1961 to 1990. Besides weather radar, there is a dense rain gauge network which comprises 49 Cassella 0.2 mm tipping-bucket rain gauges, having recording time resolution of 10 seconds (Bell and Moore, 2000). The network provides at least one rain gauge in each of the 2 km grid squares that lie entirely within the catchment. The discharge is measured at Lovington.

1.4.4   Leaf catchment

The Leaf river catchment has a 1950 km^2 area located in the north of Collins, Mississippi as shown in Figure 1-5 and its river length is 290 km in a southeastern direction from its headwaters in the southeast Scott County to its confluence with the Pascagoula River in George County (Duan et al., 2007). Leaf River watershed contains different land use types, including forest (49.9%), cropland (2.9%), pasture (22.9%), barren (15.5%), and wetlands (8.6) and the dominant land use within the watershed is forested. The mean annual precipitation is 1432 mm, and the mean annual runoff is 502 mm based on ten consecutive water-years (1951-1961) of data (daily precipitation, potential evapotranspiration estimates and observed streamflows). Leaf River Basin displays an annual cycle of six wet months, December–May, followed by six dry months, June–November. The variance of the recorded flow records peaks around February and is at a minimum from September through October. Statistics show a mean flow rate of 27.11 cm and maximum and minimum values of 1313 cm and 1.55 cm, respectively (Parrish et al., 2012)

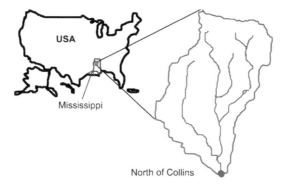

Figure 1-5. Location map of the Leaf catchment

1.4.5   Nzoia catchment

The Nzoia catchment (latitudes 1° 30'N and 0° 05'S and longitudes 34° and 35° 45'E) is located in western Kenya in the Lake Victoria basin as shown in Figure 1-6. The average annual discharge is about 1740 x 106 m^3 with the catchment area of 12,709 km^2, and a length of 334 km up to its mouth draining into Lake Victoria. The Nzoia River originates from two highland areas of Mt. Elgon and Cherengany Hills. The climate of the catchment is mainly tropical humid, with average temperatures ranging from 16°C in the highlands to 28° C in the lower semi-arid areas. The potential evapotranspiration within the catchment decreases with increasing altitude. The lowest temperature occurs in the months June to August and this coincides with the lowest evapotranspiration amounts. Generally in the drier months, the evapotranspiration exceeds rainfall amounts. The mean annual rainfall varies from a minimum of 1076 mm in the lowlands to a maximum of 2235 mm in the highlands. The area experiences four seasons in a year as a result of the inter-tropical convergence zone. There are two rainy seasons and two dry seasons, namely long rains (March to May) and the short rains (October to December). There is no distinctive dry season, but relative to the rainy seasons, the dry seasons occur in the months of January to February and in some parts, June to September. A total of 13 rainfall stations (1962-2000), 3 temperature stations (1971-2000) and 1 river gauging stations (1966-1998) were considered in this study.

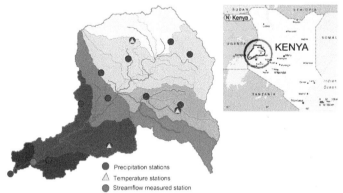

Figure 1-6. Location map of the Nzoia catchment

1.5   Terminology

Terminology related to committee modelling and uncertainty analysis of a hydrological model is presented below. These definitions may have undergone certain changes over time, due to the preferences of different authors.

Hydrological model: The hydrological modelling tool that is used to predict streamflows and their uncertainty. The lumped conceptual hydrological (rainfall runoff) model HBV is used in this thesis.

Hydraulic model: The hydrodynamic model is used for flood inundation modelling. In this study, the SOBEK 1D-2D model is used to simulate flood inundation downstream from the Nzoia catchment, Kenya.

Single optimal model: The hydrological model is calibrated by single-objective optimization.

Specialized model: A model reproducing different components of a flow hydrograph, which correspond to the characteristics of different flow regimes or models, specialized for high flows and/or low flows.

Committee models: Several models are combined using e.g., weighted averaging to provide a more comprehensive and accurate prediction. Often the combined models are specialized models.

Fuzzy committee model: A committee model where the specialized models are combined using a fuzzy membership function. The weights assigned to each specialized model's output are based on optimally designed fuzzy membership functions.

Hybrid committee model: This committee model is created by the optimal combination of conceptual and data-driven models, e.g. conceptual models specialized in high flows, and data-driven models specialized in low flows.

Predictive uncertainty model: A machine learning (data-driven) model used to encapsulate dependency of the uncertainty characteristics (*pdf* or its quantiles) on some representative variables (e.g. past events precipitation and flows). The trained machine learning models can be employed to predict the model output uncertainty, which is specific for the new hydro-meteorological situations.

SWAT-NSGAX: A particular implementation of the NSGA-II algorithm for multi-objective optimization of the SWAT model.

SWAT-SOBEK: Integration of the SWAT and SOBEK modelling systems. In this study, we estimate the uncertainty of the flood inundation assessment made by this integrated model.

Verification (or validation): Testing the model running at test data which takes place after calibration to test if the model performs on a portion of data, which was not used in calibration. The objectives of verification is to validate the model's robustness and ability to describe the catchment's hydrological response.

Cross-validation: A procedure used to minimize the overfitting in the machine learning models during their training (calibration). The common way to do this is to use the third data set (cross-validation set), apart from the training and test data sets.

Flood inundation: Consequences of excessive water in a river channel, which cause flooding and an overflow of water at the bank of the river. This information can be obtained from flood modelling and will be valuable in communicating flood risk information to decision makers, so that timely planning and mitigation measures can be taken.

1.6   Outline of the thesis

This thesis is organized into nine chapters. A brief overview of the structure is given below.

Chapter 1 introduces the problems, motivations, research questions, and objectives of the research with a description of five catchments.

Chapter 2 describes the conceptual hydrological model and techniques of computational intelligence as the major tools for data-driven modelling, including artificial neural networks, instance-based learning, and model trees. This chapter details the calibration of hydrological models, with a single objective and multi-objectives, search algorithms and the setup of HBV and SWAT hydrological models and ANN rainfall runoff models for various catchments.

Chapter 3 is devoted to multi-modelling (committees) in hydrological modelling. It starts with a brief overview of multi-models averaging, followed by different methods of model combination techniques. This chapter proposes various types of committee models, and presents the results, and their comparison, for various catchments (Alzette, Brue, Bagmati and Leaf).

Chapter 4 explores the hybrid committee of hydrological models to improve model predictions. First, it describes an overview of a hybrid model, specially for hydrological modelling, and then a calibration of a low-flow model using different objective functions. The goal was to build an ANN low-flow specialized model and a high-flow specialized model (HBV) and to find their optimal combination with appropriate membership function to form a hybrid committee model. These models are tested for the Baghmati and Leaf catchments.

Chapter 5 is devoted to parametric uncertainty analysis in hydrological modelling. It starts with a brief overview of uncertainty analysis methods and comparison of different methods of uncertainty analysis in the context of hydrological modelling. It also discusses various methods of sampling-based uncertainty analysis and their comparison results. These methods were used to analyze the uncertainty of a lumped conceptual hydrological model of the Nzoia catchment.

Chapter 6 presents the uncertainty prediction of hydrological model using machine learning techniques. Various machine learning models, such as artificial neural networks, model trees, and locally weighted regression, are tested and compared on the Bagmati and Brue catchments for the uncertainty analysis of lumped conceptual hydrological models.

Chapter 7 explores the committee of several machine learning-based predictive uncertainty models. It uses the methods for combining several predictive uncertainty models, which are built from various sampling-based uncertainty methods of hydrological modelling (also presented in Chapter 5).

Chapter 8 explores the uncertainty analysis of integrated models by linking the SWAT hydrological model and SOBEK hydrodynamic model to represent the uncertainty in flood inundation (probabilistic flood maps) in the Nzoia catchment. It also explains the setup for high performance computers for use in parallel computing for analyzing uncertainty by running multiple simulations simultaneously (in parallel).

Chapter 9 describes the conclusions of the presented research based on the various case studies included in this thesis. Finally, possible directions for further research are suggested.

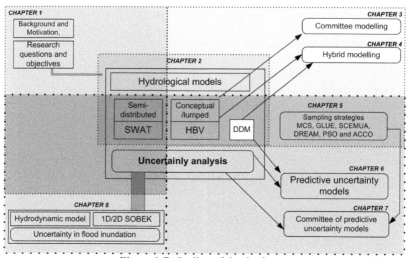

Figure 1-7. Outline of the thesis

Chapter 2
Conceptual and data-driven hydrological modelling

This chapter presents the classification of the models used in this study: hydrological models [namely, Hydrologiska Byråns Vattenbalansavdelning (HBV) conceptual hydrological model and the soil and water assessment tool (SWAT) hydrological model],and data-driven models. Their calibration by single- and multi-objective optimization is considered as well.

2.1   Introduction

Hydrological modelling tools permit to hydrologists and engineers to better understand and describe the hydrological systems in the catchments or basins. They are useful for studies of streamflows problems, water management, climate impact, and land use changes. Hydrological models represent complex, spatially and temporally distributed physical processes through straight-forward mathematical equations with parameters. These parameters can be estimated based on available knowledge, measurements of physical processes, or through calibration using input and output measurements.

The hydrological models have a variety of characteristics that require classification. Doing so ensures that the capabilities and limitations of each model can be identified correctly. The classifications are generally based on the following criteria (Singh 1995; Refsgaard 1996): (i) the extent of physical principles that are applied in the model structure; (ii) the treatment of the model inputs and parameters as a functions of space and time. In an example of first criterion, a rainfall-runoff model cab be categorized as deterministic or stochastic (refer to Figure 2-1). Deterministic models can be further categorized as physically-based or conceptual, according to the degree of complexity and physical completeness present in the formulation of the structure (Refsgaard, 1996). Figure 2-1 presents these three types of rainfall-runoff models, which include (i) data-driven models (black box), (ii) conceptual models (grey box); and (iii) physically based models (white box). If any of the input-output variables or error terms of the model are regarded as random variables having probability distribution, then the model is stochastic. An example of a stochastic model can be found in Fleming (1975), Box and Jenkins (1970), and Clarke (1973).

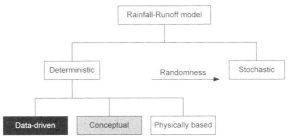

Figure 2-1. Classification of rainfall-runoff models according to physical processes (Refsgaard, 1996).

Data-driven (black box) models involve mathematical equations that are not derived from an analysis of the concurrent input and output time series in the catchments. Conceptual models are generally described as models arising from simple descriptions to equations governing relationships that aim to describe the reality (Refsgaarrd, 1997). Physically based models are built on the basis of based on the physical principles of processes wherein equations of continuity, momentum and/or energy conservation are used to describe the system behaviour.

2.2 HBV hydrological models for the considered case studies

2.2.1 HBV model brief characterization

The HBV model is a lumped conceptual hydrological model that includes conceptual numerical descriptions of the hydrological processes at the catchment scale. The model was developed at the Swedish Meteorological and Hydrological Institute (Bergström, 1976). The abbreviation HBV stands for Hydrologiska Byråns Vattenbalansavdelning (Hydrological Bureau Water Balance Section). This model has been successfully applied in all over the world (Lindström et al., 1997); its prediction uncertainty has been considered, as well (see, e.g., Seibert, 1997; Uhlenbrook et al., 1999).

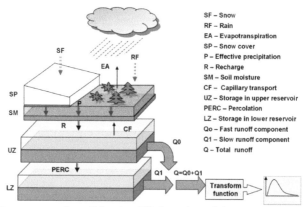

Figure 2-2. Schematic representation of the simplified version of the HBV model used in this thesis with routines for snow, soil, and runoff response (adapted from Shrestha and Solomatine, 2008).

The simplified version of the HBV model follows the structure of the HBV-96 model (Lindström et al., 1997), and its schematic diagram is shown in Figure 2-2. The model comprises subroutines for snow accumulation and melt, the soil moisture accounting procedure, routines for runoff generation, and a simple routing procedure. The snowmelt routine is based on a degree-day relation, with an altitude correction for precipitation and temperature:

$$snowmelt = CFMAX(T - TT) \qquad (2\text{-}1)$$

where TT is the threshold temperature, T is the altitude-corrected temperature; and the parameter $CFMAX$ is the melting factor. The threshold temperature is usually close to $0°C$

and is used to define the temperature above which snowmelt occurs. The threshold temperature is also used to determine whether the precipitation will fall as rain or snow. If the mean air temperature is less than the threshold temperature, precipitation is assumed to be in snow form. The snowpack is assumed to retain melt water as long as the amount does not exceed a certain fraction (given by the parameter *WHC*) of the snow. When temperature decreases below the threshold temperature, this water refreezes according to the formula:

$$refreezing meltwater = CFR \cdot CFMAX(TT - T) \qquad (2\text{-}2)$$

where *CFR* is the refreezing factor.

The soil moisture accounting routine computes the proportion of snowmelt or rainfall *P* (mm/h or mm/day) that reaches the soil surface, which is ultimately converted to runoff. This proportion is related to the soil moisture deficit and is calculated using the relation (see also Figure 2-3a):

$$\frac{R}{P} = \left(\frac{SM}{FC}\right)^{BETA} \qquad (2\text{-}3)$$

where *R* is the recharge to the upper zone (mm/h or mm/day), *SM* is the soil moisture storage (mm), *FC* is the maximum soil moisture storage (mm), and *BETA* is a parameter accounting for nonlinearity. If the soil is dry (i.e., small value of *SM*/FC), the recharge *R*, which subsequently becomes runoff, is small because the major portion of the effective precipitation *P* is used to increase the soil moisture. However, if the soil is wet, the major portion of *P* is available to increase the storage in the upper zone.

The actual evapotranspiration *EA* (mm/h or mm/day) from the soil moisture storage is calculated from the potential evapotranspiration *EP* (mm/h or mm/day) using the following formula (refer to Figure 2-3. HBV model parameters relations (a) contributions from precipitation to the soil moisture or ground water storage and (b) ratio of actual and potential evapotranspiration.Figure 2-3b):

$$EA = EP\left(\frac{SM}{FC \cdot LP}\right) \quad \text{if } SM < FC \cdot LP \qquad (2\text{-}4)$$
$$EA = EP \qquad\qquad \text{if } SM \geq FC \cdot LP$$

where *LP* is the fraction of *FC* above which the evapotranspiration reaches its potential level. The actual evapotranspiration that occurs place from the soil moisture storage depends on the soil moisture. Evapotranspiration is equal to the potential value if the relative soil moisture (i.e., *SM*/FC) is greater than *LP*. If the relative soil moisture is less than this value, the actual evapotranspiration is reduced linearly to zero for a completely dry soil.

A runoff generation routine transforms excess water *R* from the soil moisture zone to runoff. This routine consists of two conceptual reservoirs arranged vertically one over the other. The upper reservoir is a nonlinear reservoir whose outflow simulates the direct runoff component from the upper soil zone, while the lower one is a linear reservoir whose outflow simulates the base flow component of the runoff. Excess water or recharge *R* enters the upper reservoir, and its outflow is given by:

$$Q_0 = K \cdot UZ^{(1+ALFA)} \qquad\qquad (2\text{-}5)$$

where K is the recession coefficient of the upper reservoir, UZ is the storage in the upper reservoir (mm), and $ALFA$ is the parameter accounting for the non-linearity. There is also a capillary flux CF (mm/h or mm/day) from the upper reservoir to the soil moisture zone, which is calculated by the following formula:

$$CF = CFLUX\left(1 - \frac{SM}{FC}\right) \qquad\qquad (2\text{-}6)$$

where $CFLUX$ is the maximum value of capillary flux. The lower reservoir is filled by a constant percolation rate $PERC$ (mm/h or mm/day), as long as there is water in the upper reservoir. Outflow from the lower reservoir is calculated according to the following equation:

$$Q_1 = K_4 \cdot LZ \qquad\qquad (2\text{-}7)$$

where K_4 is the recession coefficient of the lower reservoir, and LZ is the storage in the lower reservoir (mm). The total runoff Q is computed as the sum of the outflows from the upper and lower reservoirs. The total runoff is then smoothed using a triangular transformation function whose base is defined by a parameter $MAXBAS$ (hours or days).

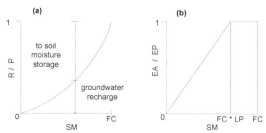

Figure 2-3. HBV model parameters relations (a) contributions from precipitation to the soil moisture or ground water storage and (b) ratio of actual and potential evapotranspiration.

2.2.2 Software development of HBV model

A component of this study effort was to develop software based on the above- mentioned model structure (HBV-96 model, Lindström et al., 1997) to ensure that the model and all its variables can be accessed in an effective way for uncertainty analysis. Preparation of input data and analysis of model results lead to increased model setup time. Such software helps to minimize this problem and furthermore enhances our understanding and prediction of the temporal dynamics of hydrologic processes.

The inputs to this model are observations of precipitation, air temperature and potential evapotranspiration. A daily time step is used for the inputs, but also possible to use a shorter time step (hourly). The evaporation values are normally monthly averages, although this software uses the daily values. Air temperature data are used for the calculations of snow accumulation and melt. They are also used to adjust potential evapotranspiration when the temperature deviates from normal values.

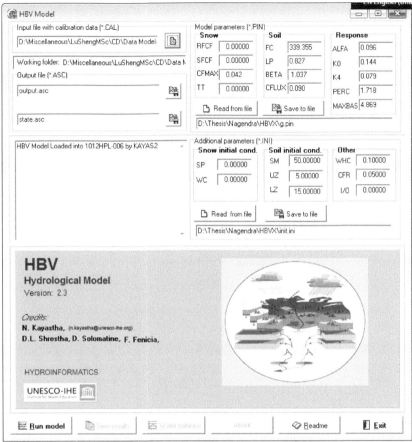

Figure 2-4. The interface of the HBV model software

The model software's interface of is designed to automatically save and retrieve hydrological data. It simulates the time series and presents the results in graphs, tables, and text files. In addition, it allows evaluation of the model by comparison of the observed and simulated streamflows and volumetric water balance in both graphical and tabular form. In addition, it allows users to easily manipulate several parameters for the purpose of manual model calibration. However, this software does not include automatic calibration as part of the model simulation. The interface allows visualization of the time series of simulated streamflows and state variables based on selected parameters. Figure 2-5 shows a time series plot of simulated and observed streamflows. The water balance plot module allows water volume difference, precipitation and evapotranspiration in a time series. The state variable module provides information regarding the soil moisture, upper zone, lower zone, percolation, fast flow distribution and slow flow distribution over time. Figure 2-4 shows the time series of the model states.

The HBV model software interface allows for:

- Developing spatially lumped conceptual hydrological models, and fitting them to data.

19

- Manually adjusting the model parameters and initial conditions.
- Simulating outputs of the model, including water balance and state variables.
- Evaluating and comparing simulated and observed streamflows.
- Summarizing performance by different measures, and displaying graphs and statistics.
- Storing the model inputs and outputs for further analysis.

This software permits a user the extraction of the model states and the visualization of their value in the time series during the comparison of observed and simulated streamflows. The behaviour of model states (based on physical rule) is often not present during calibration. The state variable plot presents the distribution of states along the time series.

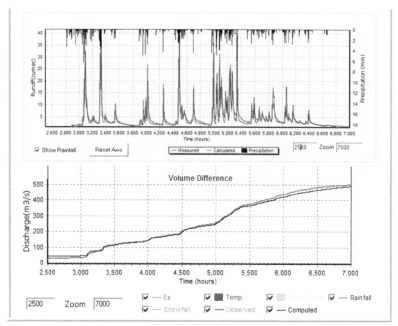

Figure 2-5. Snapshot of observed and simulated steramflows and water balance

This interface makes it possible to handle the various time scales (i. e., hourly, daily, and monthly) of input data, model states and model simulations. This model is developed in Delphi programming language and has two versions – one for command line execution and one with the visual interface. The command line execution is used for calibration and uncertainty prediction where multiple executions of the model are required. Based on the function and procedures written in Delphi, we rewrote the code in MATLAB; therefore, it can efficiently integrate the MATLAB-based algorithm with the analysis and visualization of complex multidimensional model outputs. The MATLAB version allows straightforward integration with other types of data analysis and model analysis, including model calibration and uncertainty prediction. Figure 2-4 shows a screenshot of the user control interface for selecting parameters.

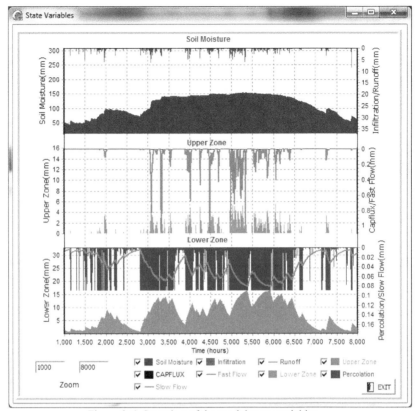

Figure 2-6. Snapshot of the model state variables

Open source development of the HBV model provides transparency, free access, and ability to modify the source code. The source code documentation is available at http://www.unesco-ihe.org/hydroinformatics/HBV/delphi_doc.html. The functions and procedures code allows for the framework to be extensible and independently developed. Moreover, the software provides ease of use, easy data development and efficient model setup and execution.

2.2.3 Models setup

The summary statistics and records of data for calibration and verification for five catchments are presented in Table 2-1. This data set covers multiple-year periods (except Brue and Alzette), all seasons, and multiple peak flows. Ideally, we must aim to split data into statistically similar sets (coverage of seasons, number and size of peaks, variance, mean, etc.). Of course, in these types of hydrological data splits, one is constrained by the requirement to maintain data in contiguous blocks (to be able to plot the time series data, such as hydrographs). Therefore, the calibration and verification data sets usually have some statistical differences.

Table 2-1. Summary of the runoff data in calibration and verification for various catchments

Statistical properties of streamflows	Data	Calibration	Verification
Alzette (Area =288 km^2)			
Period (day/month/year hour)	7/29/2000 12:00 - 8/6/2002 7:00	7/29/2000 12:00 - 8/6/2001 7:00	8/6/2001 8:00 - 8/6/2002 7:00
Number of data points	17720	8960	8760
Average (m^3/s)	4.64	5.55	3.70
Minimum(m^3/s)	0.45	0.59	0.45
Maximum (m^3/s)	51.41	51.41	31.15
Standard deviation(m^3/s)	5.35	5.52	5.00
Bagmati (Area=3500 km^2)			
Period (day/month/year)	01/01/1988 - 31/12/1995	01/03/1991 - 30/121995	01/07/1988 - 28/02/1991
Number of data	2922	1767	1155
Average (m^3/s)	150	150.8	148.6
Minimum (m^3/s)	5.1	5.1	7.7
Maximum (m^3/s)	5030	5030	3040
Standard deviation(m^3/s)	271.2	280.5	256.4
Brue (Area=135 km^2)			
Period (day/month/year hour)	24/06/1994 05:00 – 31/05/1996 13:00	24/06/1994 05:00 – 24/06/1995 04:00	24/06/1995 05:00 – 31/05/1996 13:00
Number of data	16977	8760	8217
Average (m^3/s)	1.91	2.25	1.53
Minimum (m^3/s)	0.15	0.17	0.14
Maximum (m^3/s)	39.58	39.58	29.56
Standard deviation(m^3/s)	3.14	3.68	2.37
Leaf (Area=1924 km^2)			
Period (day/month/year)	28/07/1951 - 21/09/1961	28/07/1951 - 25/07/1957	26/07/1957 - 21/09/1967
Number of data	3717	2190	1527
Average (m^3/s)	28.28	23.02	35.81
Minimum (m^3/s)	1.56	1.56	2.92
Maximum (m^3/s)	1313.91	549.35	1313.91
Standard deviation(m^3/s)	64.48	47.37	82.51
Nzoia (Area=12709 km^2)			
Period (day/month/year)	01/01/1970 - 12/12/1985	01/01/1970 - 12/12/1979	01/01/1980- 12/12/1985
Number of data	5675.00	3544.00	2131.00
Average (m^3/s)	88.50	95.88	76.24
Minimum (m^3/s)	8.62	15.36	8.62
Maximum (m^3/s)	578.68	578.68	381.49
Standard deviation(m^3/s)	63.85	65.65	58.74

2.2.3.1. HBV model setup for the Brue catchment

The hourly data involving rainfall, discharge, and automatic weather data (temperature, wind, solar radiation, etc.) were computed from the 15-minute data. The basin average rainfall data were used in the study. The hourly potential evapotranspiration was computed using the modified Penman method recommended by the United Nations Food and Agriculture Organization (FAO) (Allen et al., 1998). One-year hourly data from 1994/06/24 05:00 to 1995/06/24 04:00 was selected for calibration of the HBV hydrological model and data from 1995/06/24 05:00 to 1996/05/31 13:00 was used for the validation (testing) of the hydrological model. Each of the two data sets represents almost a full year of observations, and their statistical properties are shown in Table 2-1.

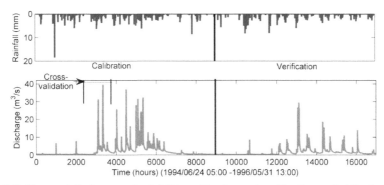

Figure 2-7. Observed discharge and rainfall in calibration and verification period for the Brue catchment

Table 2-2. Ranges and calibrated values of the HBV model parameters for Brue catchment.

Parameter	Description and unit	Ranges	Calibrated value
FC	Maximum soil moisture content (mm)	100-300	160.335
LP	Ratio for potential evapotranspiration (-)	0.5-0.99	0.527
ALFA	Response box parameter (-)	0-4	1.54
BETA	Exponential parameter in soil routine (-)	0.9-2	1.963
K	Recession coefficient for upper tank (/hour)	0.0005-0.1	0.001
K4	Recession coefficient for lower tank (/hour)	0.0001-0.005	0.004
PERC	Maximum flow from upper to lower tank (mm/hour)	0.01-0.09	0.089
CFLUX	Maximum value of capillary flow (mm/hour)	0.01-0.05	0.0038
MAXBAS	Transfer function parameter (hour)	8-15	12

The HBV model has 13 parameters (4 parameters for snow, 4 for soil, and 5 for the response routine). Because there is little or no snowfall in the catchment, the snow routine was excluded and only 9 parameters (Table 2-2) were used. The model was first calibrated using the global optimization routine – adaptive cluster covering algorithm, ACCO (Solomatine, 1999) to find the best set of parameters. Subsequently, manual adjustments of the parameters were made by visual comparison of the observed and simulated hydrographs. The ranges of parameter values for calibration and uncertainty analysis were established based on the ranges of calibrated values from the other model applications (e.g., Braun and Renner, 1992) and the hydrologic knowledge of the catchment. The ranges were extended when the solutions were found near the border of the parameter ranges and re-calibration of the model was performed with the extended range of the parameters.

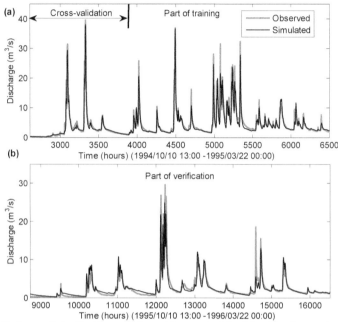

Figure 2-8. Hydrograph for the Brue catchment in a part of the (a) calibration period and (b) verification period.

The model was calibrated using the Nash and Sutcliffe efficiency (*NSE*) value (Nash and Sutcliffe efficiency, 1970) as a performance measure of the HBV model. An *NSE* value of 0.96 was obtained for the calibration period. The model was validated by simulating the flows for the independent verification data set, and the *NSE* is 0.83 for this period. Figure 2-8 shows the observed and simulated hydrograph in a part of calibration and in verification period. HBV model is quite accurate, but its error (uncertainty) is quite high during the peak flows.

2.2.3.2.   HBV model setup for the Bagmati catchment

The long-term mean annual discharge of the river at the station was 151 m^3/s but the annual discharge varied from 96.8 m^3/s in 1977 to 252.3 m^3/s in 1987 (Department of Hydrology and Meteorology, 1998). The daily potential evapotranspiration was computed using the modified Penman method recommended by FAO (Allen et al., 1998). A total of 1767 numbers of daily records from 1 March 1991 to 31 December 1995 were selected for calibration of the process model (in this study the HBV hydrological model). Data from 1

January 1988 to 28 February 1991 were used for the validation (verification) of the process model. The first two months from 1 March 1991 to 29 April 1991 of the calibration data were used as a warming-up period and hence excluded during calibration. The separation of the 8 years of data into calibration and validation was done on the basis of hydrological seasons. The statistical properties of the runoff data are presented in Table 2-1

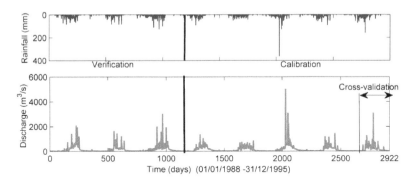

Figure 2-9. Observed discharge and rainfall in calibration and verification period for the Bagmati catchment

The *NSE* value of 0.87 was obtained for the calibration period; this value corresponds to the root-mean-squared error (*RMSE*) value of 102.63 m³/s. The *NSE* was 0.83 for this period with the RMSE value of 111.56 m³/s. Please note that the standard deviation of the observed discharge in the validation period is 9% lower than that in the calibration period. The uniform ranges of parameters (Table 2-3) are used for calibration of the HBV model using the ACCO algorithm. The observed and simulated discharges in the verification period are shown in Figure 2-25, and their performances are presented in Table 2-4.

Table 2-3. Ranges and optimal values of the HBV model parameters for Bagmati catchment.

Parameter	Description and unit	Ranges	Value
FC	Maximum soil moisture content (mm)	50 - 500	354.98
LP	Limit for potential evapotranspiration	0.3 - 1	0.71
ALFA	Response box parameter	0 - 4	0.167
BETA	Exponential parameter in soil routine	1- 6	1.0002
K	Recession coefficient for upper tank (/day)	0.05 - 0.5	0.280
K4	Recession coefficient for lower tank (/day)	0.01 - 0.5	0.0767
PERC	Maximum flow from upper to lower tank (mm/day)	0 - 8	7.99
CFLUX	Maximum value of capillary flow (mm/day)	0 - 1	0.00006
MAXBAS	Transfer function parameter (day)	1 - 3	2.546

2.2.3.3. HBV model setup for the Nzoia catchment

The HBV model with 9 model parameters was configured for calibration with the data period from 01/01/1970 to 31/12/1979 and verification period from 01/01/1980 to 31/12/1985, with these data sets consisting of 3544 and 2131 daily data respectively. The first two months of data were considered as the model warm-up period. The daily potential evapotranspiration was computed using the modified Penman method. Each of the two data sets represents a different duration of observations. Their statistical properties are given in Table 2-1 and rainfall and observed runoff are shown in Figure 2-10. The observed and simulated discharges are shown in Figure 5-15 and their performances are presented in Table 5-3 in Chapter 5.

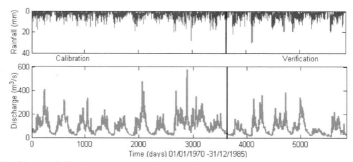

Figure 2-10. Observed discharge and rainfall in calibration and verification period for the Nzoia catchment

2.2.3.4. HBV model setup for the Leaf catchment

The daily data period is from 28/07/1951 to 25/07/1957 and verification 26/07/1957 to 21/09/1967 which include 2190 and 1527 daily data respectively. The first two months of data were considered as the model warm up period. The statistical properties of two data sets are shown in Table 2-1 and the observed rainfall and discharge are shown in Figure 2-11. The observed and simulated discharges by HBV in verification period are shown in Figure 2-26 and their performance are presented in Table 2-5.

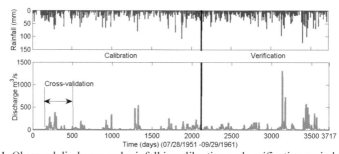

Figure 2-11. Observed discharge and rainfall in calibration and verification period for the Leaf catchment

2.2.3.5. HBV model setup for the Alzette catchment

The hourly basin average rainfall data and discharge was used for the Alzette catchment. One-year hourly data from 7/29/2000 12:00h to 8/6/2001 00:00h were selected for calibration, and data from 8/6/2001 8:00h to 8/6/2002 7:00h were used for the verifying (testing) of the hydrological model (Table 2-1). The observed rainfall and discharge are shown in Figure 2-12. The fragment of observed and simulated discharges by HBV in the verification period is shown in Figure 3-6 (a) and their performances are presented in Table 3-6 in Chapter 3.

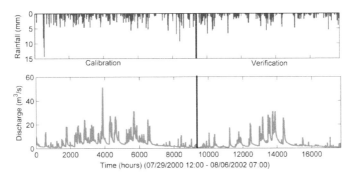

Figure 2-12. Observed discharge and rainfall in calibration and verification period for Alzette catchment

2.3 SWAT model for the Nzoia catchment

2.3.1 SWAT model description

The soil and water assessment tool (SWAT, Arnold et al., 1998) is a spatially-distributed and semi-physically-based hydrological model that is used for the modelling of river basins, catchments and watersheds. SWAT incorporates physical processes and lumps together parts of catchments that have the same soil and land cover properties. The main processes in the model are precipitation, evapotranspiration, runoff, groundwater flow, and storage and their interaction within the catchment. The model also accounts for vegetation growth and management practices occurring in the catchment (Figure 2-13). The physical processes associated with water movement, sediment movement, crop growth, and nutrient cycling can be directly modelled by SWAT using input data (Arnold et al., 2005). The SWAT model requires topography, soil, land use and weather data whereby the watershed is divided into a number of sub-catchments using topography data. The sub-catchments-based simulation is particularly beneficial when different areas of the catchment are dominated by land uses or soil pattern in which these properties directly impact hydrology. Each sub-catchment is subsequently again subdivided into Hydrologic Response Units (HRUs; unique combination of land use, and soil). Inputs for each sub-catchment are grouped into the following categories: climate, HRUs, ponds/wetlands, groundwater, and the main channel (or reach) draining the sub-catchment.

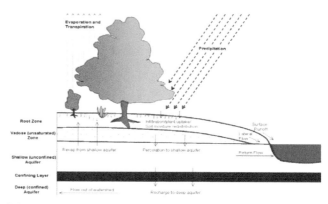

Figure 2-13. Schematic representation of hydrologic cycle in SWAT (Neitsch et al., 2005)

The driving force behind SWAT is water balance, which is determined by calculating the input, output, and storage changes of water at the catchment's surface. The hydrologic cycle as simulated by SWAT is based on the water balance equation:

$$SW_t = SW_0 + \sum_{i=0}^{t}(R_{dry} - Q_{surf} - E_a - w_{seep} - Q_{gw})$$ (2-8)

where SW_t is the final soil water content, SW_0 is the initial soil water content on day i, t is the time (days), R_{day} is the amount of precipitation on day i, Q_{surf} is the amount of surface runoff on day i, E_a is the amount of evapotranspiration on day i, W_{seep} is the amount of water entering the vadose zone from the soil profile on day i, Q_{gw} is the amount of return flow on day i.

The subdivision of the watershed enables the model to reflect differences in evapotranspiration for various crops and soils. Runoff is predicted separately for each HRU and routed to obtain the total runoff for the watershed, which provides a much better physical description of the water balance. In SWAT the water balance of each HRU in SWAT is represented by four storage volumes: snow, soil profile (0–2 m), shallow aquifer (typically 2–20 m), and deep aquifer. Once water is introduced into the system as precipitation, the available energy (specifically solar radiation) exerts a major control on the movement of water in the land phase of the hydrologic cycle. These processes are significantly affected by temperature and solar radiation and include snow fall, snow melt and evaporation. Because evaporation is the primary water removal mechanism in the watershed, the energy inputs become very important in reproducing or simulating an accurate water balance.

2.3.2 Inputs for the SWAT model

Precipitation is one of the influential inputs for runoff simulation and the degradation of rain gauge networks strongly affects the simulated hydrographs (e.g., Sun et al. 2002). Generally climate data (e. g., precipitation and temperature) are measured at the individual gauge stations in the field and used as input data for hydrological models. In a distributed (or semi-distributed) hydrological model, input data should be fed into sub-catchment models to cover spatially the entire area of the catchment. Hence, spatial interpolation of climate data is necessary to upscale the data from point to the sub-catchment area. However, interpolation is possible only if the rain gauge data is available for the area of interest. Various interpolation methods exist in hydrological modelling, for example, the Thiessen polygon method (Thiessen, 1911), inverse distance weighting (e.g. Shepard, 1968), and geostatistical methods (kriging) (Goovaerts, 1997).

Tabios and Salas (1985) presented a comparison of various interpolation methods, namely, kriging, Thiessen polygons, inverse distance weighting (IDW), polynomial trend surfaces, and inverse square distance algorithms. They found that the Kriging method is the most effective method among the reviewed interpolation methods for rainfall. We used the ordinary Kriging method for interpolation of rainfall and temperature for the Nzoia catchment. The spatial interpolation of precipitation for one event (one day) is shown in Figure 2-15.

Figure 2-14 Rain gauge and temperature stations, red circles indicate rain gauges, blue triangles indicate temperature stations

Figure 2-15. The spatial distribution of precipitation at an event of 13 November 1990 over the Nzoia, as derived using the ordinary Kriging interpolation method, legend is at scale of by x10 mm (Nzoia catchment)

2.4   Calibration of hydrological models

Hydrological models contain several parameters that cannot be directly measured. The process of estimating the parameters is called "model calibration". Generally, calibration is carried out by optimizing the model parameters so that the model output matches as closely as possible the observed responses of the hydrological system over any historical period of time (Figure 2-16). There are two calibration approaches: manual and automatic. Manual adjustment of the parameter values is labour intensive, and its success strongly depends on the experience of the modeller. Because of the time-consuming nature of manual calibration, there has been a great deal of research into the development of automatic calibration methods (e.g., Duan et al., 1992; Yapo et al., 1996; Solomatine et al., 1999; Madsen, 2000; Vrugt et al, 2003). Automatic calibration is quite easy to implement in hydrological modelling; however, the quality of optimization depends on the objective function choice and data quality and efficiency of the calibration approach.

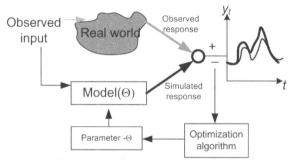

Figure 2-16. Calibration of model.

Optimization algorithms are generally classified into two categories, local search and global search. The difficulties associated with the local searches include the high degree of nonlinearity of the response surface, the existence of multiple local optima in the search space and the discontinuities of first and second order derivatives. These problems inspired the use of global optimization algorithms (typically, randomized search) that are capable of carrying out effective searches regardless of the response surface.

Randomized search algorithms aim to minimize differences between selected features of modelled and observed streamflows by systematic trial alterations in the values of the model parameters. The objective function, i.e., the quantitative measure of the fit of modelled runoff to the observed runoff, is calculated after each parameter alteration. Successful alterations are those that cause a reduction in the value of the objective function. During the search only the parameter set associated with the current lowest objective function value is retained. At the end of a search, this lowest value is regarded as the optimal (best) parameter set (or several parameter sets also retained as well to represent other "good" models).

2.4.1   Single objective optimization

A single objective optimization aims to find the "best" solution that reaches a target value (minimum or maximum) of a single objective function (which may lump several objectives into one). The optimization algorithms require the definition of an objective function and the ranges for the decision variables. These algorithms can be found in many publications; among them most popular algorithms that include the shuffled complex evolution method developed at the University of Alabama (SCE-UA, Duan et al., 1992), the genetic algorithm (e.g. Wang 1991), Adaptive cluster covering (ACCO, Solomatine, 1999) and others. SCE-UA algorithm combines the strength of the simplex downhill method, controlled random search and concept of shuffling. The robustness and efficiency of this algorithm have been evaluated by a number of studies (e.g. Duan et al 1992). In this thesis, the single objective optimization algorithm ACCO algorithm is used. This algorithm uses clustering as the first step, but it is followed by a global randomized search, rather than local search. In general, it utilizes a combination of accepted ideas including reduction, clustering, and covering. Its main principles are outlined below and a detailed description of the algorithm can be found in Solomatine (1999).

1. Clustering (i. e.,identification of groups of mutually close points in search space) is used to identify the most promising sub-domains in which to continue the global search by active space covering (single-extremum search in each region).

2. Covering: Each sub-domain is covered randomly. The values of the objective function are then assessed at the points drawn from the uniform distribution or some other distribution. Covering is repeated multiple times and each time the subdomain is progressively reduced in size.

3. Adaptation: Adaptive algorithms update their algorithmic behaviour depending on the new information revealed about the problem under consideration. In ACCO, adaptation is done by shifting the subregion of search, shrinking it, and changing the density (number of points) of each covering - depending on the previous assessments of the global minimizer.

4. Periodic randomization: Due to the probabilistic character of point's generation, any randomized search strategy may simply miss a promising region for search. In order to reduce this risk, the initial population can be re-randomized, (i.e., the problem is solved several times). Depending on the implementation of each of these principles, it is possible to generate a family of various algorithms that are suitable for certain situations. (e.g. with non-rectangular domains (hulls), non-uniform sampling and with various versions of cluster generation and stopping criteria).

2.4.2   Multi objective optimization

A single objective function is often inadequate for properly measuring the simulation of all the important characteristics of a system in calibration of a catchment model (van Griensven and Bauwens, 2003), possibly resulting in a loss of information. Futhermore, when moving from one solution to another, it can be difficult to deal with conflicting objectives and the choice of objectives. Gupta et al. (1998) pointed out that there might not exist "statistically correct" choice for the objective function therefore no statistically correct "optimal choice" for the model parameters. Furthermore, reorganization of the multi-objective nature of the calibration problem and recent advances in computational power have led to more complex hydrological models, often predicting multiple hydrological fluxes simultaneously. These issues have spurred increasing interest in the multi-objective calibration of hydrological model parameters (Gupta et al., 1998; Yapo et al., 1998; Madsen, 2000; Khu and Madsen, 2005).

The description of the multi-objective calibration problem is stated as follows:

$$\min\{f_1(\theta),...,f_m(\theta)\} \tag{2-9}$$

where m is the number of objective functions, $f_i(\theta)$, $i = 1, ..., m$ are the individual objective functions, and θ is the set of model parameters to be calibrated. Due to trade-offs between the different objectives, the solution to Equation (2-9) will no longer, in general, be a single unique parameter set. Instead, there exist several solutions that constitute a so-called Pareto-optimal set or non-dominated solutions set. Any solution θ^* belongs to the Pareto set when there is no feasible solution θ that will improve some objective values without degrading performance in at least one other objective. Mathematically, the solution θ^* is Pareto-optimal (i) if and only if $f_i(\theta^*) \leq f_i(\theta)$ for all $i = 1, ..., m$ and (ii) $f_j(\theta^*) < f_j(\theta)$ for some $j = 1, ..., m$. According to these two statements, the Pareto-optimal solution θ^* has at least one smaller objective value compared with any other feasible solution θ in the decision space, while performing as well or worse than θ in all remaining objectives.

The multiobjective optimisation algorithm used in this study is the non-dominated sorting genetic algorithm (NSGA-II) proposed by Deb et al. (2002). NSGA-II is capable of handling a large number of objective functions and provides an approximate representation of the Pareto set with a single optimisation run. The NSGA-II algorithm is outlined briefly as follows:

1. Generate an initial population of size p randomly in the domain of the feasible range.

2. Evaluate the population and sort it based on non-domination using a bookkeeping procedure to reduce the order of the computation.

3. Classify the population into several Pareto fronts based on the non-domination level. Individuals belonging to the first Pareto front are assigned with rank 1, individuals belonging to the second Pareto front (the second Pareto front is the Pareto front after removing the individuals from the first front) are assigned with rank 2, and so on.

4. Form a new population by generating an offspring population from the parents and combining them with the parents (following standard GA procedure).

5. Compute the crowding distance for each individual.

6. Select the individuals based on a non-domination rank. The crowding distance comparison is made if the individual belongs to the same rank.

7. Repeat the steps 3–6 until the stopping criterion is satisfied. The stopping criterion may be a specified number of generations, maximum number of function evaluations or computation time.

There is a necessity for well-organized multi-objective calibration procedures that are capable of exploiting all of the useful information and important characteristics about the physical system of the catchment. The objectives to be optimized can be the goodness-of-fit estimators (e.g., coefficient of efficiency, root mean squared error, and residuals), multiple variables (e.g., streamflows, sediment, and nutrients, etc.), and multiple sites (multiple locations of observations). Such an approach allows the generation of the so-called Pareto front – a set of optimal solutions that represent the trade-offs between the objectives. The methodology for solving the multiple-objective global optimization problem is presented in many hydrology-related papers; the MOCOM-UA algorithm, (Yapo et al., 1998), MOSCEM-

UA global optimization algorithm (Vrugt et al., 2003), AMALGAM (Vrugt and Robinson, 2007), SPEA2 (Zhang et al., 2008), MEAS (Efstratiadis and Koutsoyiannis, 2008), Genetic algorithms (Khu and Madsen 2005; Tang et al., 2006; Shafii and Smedt 2009) and MO-ROPE (Krauße et al.,2011).

In particular, several researchers have comprehensively demonstrated the multi-objective genetic algorithm (NSGA-II) in the context of SWAT modelling (e. g. Remegio et al., 2007; Bekele et al., 2007; Maringanti et al., 2009; Dumedah et al., 2010). Remegio et al. (2007) successfully applied the NSGA-II and Pareto ordering optimization in the automatic calibration of SWAT for daily streamflows. Bekele et al. (2007) used NSGA-II for two scenarios; in the first, specific objective functions to fit different portions of the time series, and in the second, the calibration was performed using data from multiple gauging stations. Maringanti et al. (2009) applied NSGA-II for the selection and placement of best management practices for nonpoint source pollution control, where total pollutant load from the watershed and net cost increase from the baseline were the two objective functions minimized during the optimization process. Dumedah et al. (2010) outlined an automated framework using the distribution of solutions in both objective space and parameter space to select solutions with unique properties from an incomparable set of solutions. However, these ample demonstrations are presented only in papers, the developed tools are not directly accessible and they are often difficult to set up. For these reasons, we developed SWAT-NSGAX tool to calibrate the SWAT model in the presence of multiple objectives.

NSGA-II is one of the efficient, multi-objective evolutionary algorithms. It has been widely applied in various disciplines. Some of the recent applications can be found in the literature, e.g. for the optimization of water distribution networks (Atiquzzaman et al., 2006, Herstein and Filion, 2011), operations of a multi-reservoir system (Chang and Chang 2009), and rehabilitation of urban drainage systems (Bareto et al., 2010). The complete description of this algorithm can be found in Deb et al. (2002).

2.4.3   SWAT-NSGAX tool and its application

Kayastha et al. (2011) developed the SWAT-NSGAX tool that link the SWAT model and the NSGAX (Barreto et al., 2009, 2010) (implementing the NSGA-II algorithm) using the SWAT-CUP model standard. The tool allows users to perform an automated calibration that considers multiple objective functions in an order to compute the Pareto-front. Doing so gives a set of optimal solutions that represent the trade-offs between the objectives, which are often due to underlying model-structural errors.

In order to generate a solution space for each model parameter in SWAT-NSGAX, the user should first select the upper and lower bounds for the SWAT parameters and provide the values for the NSGA-II operational parameters (e.g., population size, number of generations, the probability of crossover, and mutation value). The calibration process is intended to minimize the identified objective functions. Based on the generated initial population each newly sampled parameter set is sent to SWAT, where the model parameter values are changed before producing a new model output. The simulated outputs for the specified parameter set are used to compute its goodness-to-fit (e.g., coefficient of efficiency and root mean squared error). This process is repeated for all parameter sets in that population and non-dominated sorting is performed to sort parameter sets into different fronts. Selection and reproduction operations are performed on the population followed by crossover and mutation to create a new population of parameter sets for evaluation. The generation of the new population incorporates crowding operators by distributing solutions to less-crowded areas in order to provide range along with selected solutions. The process is repeated until the difference between the fitness values for the previous and current populations is below a user-defined (low) value.

The generated Pareto front curves from NSGAX are simultaneously visualized in a graphical interface and saved in text files. Additionally the NSGAX stores files according to the user-defined project name. These files contain the data of the best and of all populations (parameter sets) with their corresponding objective function values. All of these values could be further used for uncertainty analysis, as well as for a cluster analysis to evaluate the distribution of solutions in the objective and parameter spaces.

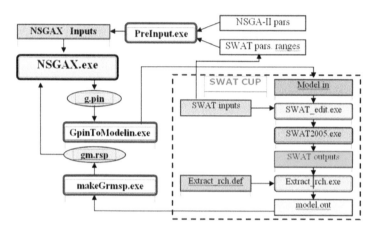

Figure 2-17. Linking between NSGAX and SWAT-CUP

The schematic diagram of the dynamic linking of SWAT and NSGAX is presented in Figure 2-17. In this framework, three programs were mainly used to couple SWAT and NSGAX: *PreInput.exe*, *GpinToModelin.exe* and *makeGrmsp.exe*; they were written in the Delphi programming language. The inputs for NSGAX are prepares by the program *PreInput.exe*, while *GpinToModelin.exe* reads data from file 'g.pin' which is generated by NSGAX and updates it in the SWAT-CUP input file 'model.in'. The program *makeGrmsp.exe* reads the output file from SWAT's 'model.out' and evaluates the user-defined objective function value, and updates it into the NSGAX input file 'gm.rsp'.

This tool was successfully used to solve a multi-objective calibration problem for (i) flow and pesticides in a catchment in Nil, Belgium (Kayastha et al., 2011) and (ii) flow and sediment routing in a catchment in Lillebæk, Denmark (Lu et al., 2012). The generated Pareto-fronts allow for identifying trade-offs between different parameterized (flow, sediment and pesticide parameters) models.

Kayastha et al. (2011) used this tool for the SWAT model of pesticides transport with an objective of optimizing the streamflows and pesticides in the Nil River in the central part of Belgium. *NSE* was used to evaluate model performance. First multi objective calibration was carried out for the flow and pesticide observations by sampling of the flow parameters, second by sampling both flow and the pesticide parameters, and third with and without point of sources. The performance of different model variations was assessed by using the resulting Pareto fronts.

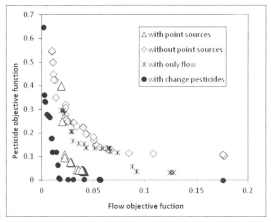

Figure 2-18. Pareto plots generated by the NSGAX-SWAT tool (Kayastha et al 2011)

Lu et al. (2012) used this tool to calibrate two different rivers sediment routing methods in the SWAT model for the Lillebæk catchment, southeast coast of the Island Fyn, Denmark. They plotted the Pareto of the two objectives of which were calculated with (i) a routing method with the simplified Bagnold equation (default) and (ii) a routing method that separated the river bank and river bed routing. (This experiment used two versions of the models – SWAT2005 and SWAT2009. Their results showed that the Pareto front can be better represented using SWAT2009 than SWAT2005.)

2.5 Data driven modelling

2.5.1 Introduction

Data-driven modelling utilizes the theoretical foundations of machine learning (Mitchell 1997) and produces a mathematical model of the relationship between the input and output even when the underlying mechanism is unknown or hard to describe. This type of modelling can learning from data without requiring prior knowledge of the model component process.

There are three main types of machine learning methods: supervised, unsupervised and reinforcement. Supervised learning is a machine learning technique for learning functional relationships between the input data and the output from the training data. The training data consist of pairs of input data and desired outputs. The output of the function can be a continuous value (in this case the style of learning is called regression), or it can predict a class label of the input variable (this style of learning is called classification). The task of a supervised learner is to predict the output for any new input vector after having seen a number of training examples. Examples of supervised learning methods are artificial neural networks, decision trees, model trees, instance-based learning, support vector machines, etc. In unsupervised learning, the training data consist of only input data and the target output is unknown. One form of unsupervised learning is clustering, where the data are grouped into subsets according to their similarities.

Machine learning methods can also be classified into eager learning or lazy learning. The former type constructs explicitly constructs the model to predict or classify the output as soon as the training examples are presented. The latter type simply stores the presented training data and postpones using it until a new instance or input has to be predicted or classified. Examples of eager learning are artificial neural networks, decision trees, model trees, etc. Lazy learning methods include k-Nearest Neighbour and locally weighted regressions.

This study uses supervised learning techniques (i. e., artificial neural networks and model tress), and a lazy learning technique (locally weighted regression).

Machine learning principle

Learning (or training) is the process of minimizing the difference between observed response y and model response $\hat{y}$ through the optimization procedure as shown in Figure 2-19. Learning is an iterative procedure that starts from the initial guess of the parameter vector **w** and updates it by comparing the models' response $\hat{y}$ with the observed response y.

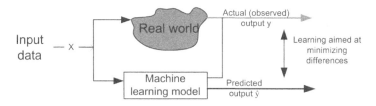

Figure 2-19. The learning process of a machine learning algorithm

The machine learning algorithm learns an unknown mapping between predicted and observed data. Learning is done on the training data and the remaining data are used to check the generalizability of the trained model. During the learning phase (training), the prediction error $y-\hat{y}$ is used to update the machine learning model. The predictive performance model is evaluated by presenting the unseen input data (verification or validation) data.

2.5.2 Machine learning in data-driven rainfall-runoff modelling

Machine learning techniques have been used extensively in the area of rainfall-runoff modelling. They are used to improve the accuracy of prediction/forecasting made by process-based rainfall-runoff model. Among machine learning techniques the artificial neural networks (ANNs) are the most popular and widely used technique. Maier and Dandy (2000, 2010) provide an extensive review of the application of ANNs in hydrological modelling. Apart from these, ANNs are also used with a combination of process-based models in flow simulations (e.g. Corzo et al. 2009). Shrestha et al. (2009; 2013) proposed to estimate the uncertainty of the runoff prediction by hydrological models (instead of the runoff itself) using machine learning techniques, which are described in Chapter 6.

2.5.3 Artificial neural networks

ANNs are artificial intelligence-based information processing tools inspired by biological processes of a human brain. They can be useful when large search spaces of human expertise are required. ANNs resemble the brain processor in two respects; a neural network acquires knowledge through learning processes, and inter-neuron connection strengths known as synaptic weights, are used to store the knowledge (Haykin, 1999). ANNs have the ability to represent both linear and complex, nonlinear relationships between inputs and outputs. They also have the ability to learn these relationships directly from the data being modelled. Hence, ANNs do not need detailed knowledge of the internal system. They have been successfully applied in fields for control, classification, pattern recognition, dynamic systems modelling, and time series forecasting.

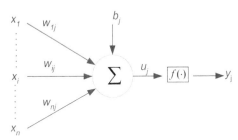

Figure 2-20. Schematic diagram of an artificial neuron.

ANNs consist of a large number of simple processing elements called *neurons* or *nodes*. Each neuron is connected to other neurons by means of direct links, each being associated with a weight that represents information being used by the network in its effort to solve the problem. The neural network can be in general characterized by its architecture (the patterns of connection between the neurons), its training or learning algorithms (the methods of determining the weights on the connections) and its activation functions. The architecture of a typical neural network with a single neuron is shown in Figure 2-20. It consists of five basic elements: (i) input nodes for receiving input signals x_1, ..., x_p, (ii) a set of connecting links (*synapses*), each of which is characterised by a weight w_{ij}, (iii) aggregating function to sum the input signals, (iv) an activation function that calculates the activation level of the neuron; and (v) output nodes y_1, ..., y_l.

The processing of each neuron is carried out in two steps: (i) summing of the weighted input signals, and (ii) applying an activation function to the sum for limiting the amplitude of the output of a neuron. Mathematically, this process described by the following two equations:

$$u_j = \sum_{i=1}^{p} w_{ij} x_i \qquad\qquad (2\text{-}10)$$

$$y_j = f(u_j + b_j) \qquad\qquad (2\text{-}11)$$

where w_{ij} is the weight connecting the input i to the neuron j. The effective incoming signal u_j and bias b_j is passed through activation function $f(.)$ to produce the output signal y_j. The main difference between the commonly used neurons lies in the type of the activation function. Their functional form determines the response of a node to the total input signal; however, these activation functions have one thing in common: they all restrict the input signals to certain limits. Some commonly used activation functions are linear, binary, sigmoid and tangent hyperbolic.

The measured data is used to build an ANNs model where the data are divided into three sets: training, validation, and test sets. (This is actually true for any machine learning method.) Training is a set of examples used for learning – that is, to fit the parameters (weights) of the ANN. Validation is a set of examples used to tune the parameters of an ANN – for example, to choose the number of hidden units in a neural network. Test data is a set of examples used only to assess the performance of a fully-specified model. First the ANN model is trained to represent the relationships and processes within the training and validation data set. Once the model is adequately trained, it is able to generalize and calculate the relevant output for the set of input data. This output is subsequently compared with the measured test data set. The model is considered to behave satisfactorily if its performance during the testing period is similar to that during the training period (or at least does not differ too much).

A multi-layer perceptron (MLP) consists of an input layer, an output layers, and at least one intermediate layer between the input and output layers. The first layer is the input layer, and receives the input signal. The intermediate layers are known as hidden layers and do not have direct connection to the outer world. The last layer is the output layer at which the overall mapping of the network input is made available and thus represents the model output. The nodes in one layer are connected to those in the next, but not to those in the same layer. Thus, the information or signal flow in the network is restricted to a layer by layer flow from the input to output through hidden layers. Figure 2-21 shows an example of a three-layer MLP with one hidden layer.

Training (i.e. computing the weights) of the MLP networks is done with the back-propagation algorithm which is the most popular algorithm capable of capturing a variety of nonlinear error surfaces. A gradient descent technique minimises the network error function. The back propagation algorithm involves two steps. The first step is feed forward pass, in which the input vectors of all the training examples are fed to the network and the output vectors are calculated.

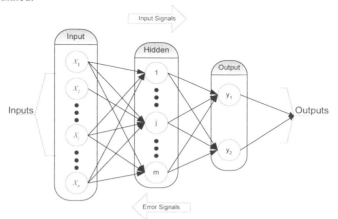

Figure 2-21. A three-layer MLP architecture.

The performance of the network is evaluated using an error function based on the target and network outputs. After the error is computed, the back propagation step starts, in which the error is propagated back to adjust the network weights and bias. The iteration continues until the outputs of the network match the targets with a desired degree of accuracy. In a typical application, the weight update loop in back propagation may be iterated thousands of times. A variety of termination conditions can be used to halt the iteration. One may choose to halt after a fixed number of iterations once the error on the training examples falls below some threshold or once the error on a separate validation set of examples meets some criterion. The choice of termination or stopping criterion is important and has been discussed by many authors (see, e.g., Haykin 1999).

The back propagation method does not guarantee convergence to an optimal solution. However, because local minima may exist, it appears in practice it appears to solutions in almost every case. In fact, standard multi-layer, feed-forward networks with only one hidden layer have been found capable of approximating any measurable function to any desired degree of accuracy. Detailed description of the back propagation algorithm can be found in Haykin (1999) and Principe et al. (1999) among others.

2.5.4 Model trees

Model trees (or M5 model trees) (Quinlan (1992); see also Witten and Frank (2000)) are the machine learning technique equivalent to a piece-wise linear regression model. They use the 'hard' (i.e. yes–no) splits of input space into regions progressively narrowing the regions of the input space. Thus model trees are a hierarchical (or tree-like) modular models that have splitting rules in non-terminal nodes and the regression models at the leaves of the tree built for the non-intersecting data subsets. Once these models are formed in the leaves of the tree, then prediction with the new input vector consists of the following two steps: (i) classifying the input vector to one of the subspace by following the rules in the non-leaf nodes of the tree; and (ii) running the corresponding regression the model in the leaf node. A more formal description of model tree algorithm is presented below.

Assume we are given a set of N data pairs $\{\mathbf{x}_i, y_i\}$, $i = 1, \ldots, n$, denoted by D. Here $\mathbf{x}$ is p dimensional input vector (i.e. $x^1, \ldots, x^p$) and y is the target. Thus, a pair of input vector and target value constitute the example, and the aim of building model tree is to map the input vector to the corresponding target by generating simple linear equations at the leaves of the trees. The first step in building a model tree is to determine which input variable (often called *attribute*) is the best to split the training D. The splitting criterion (i.e., the selection of the input variable and splitting value of the input variable) is based on treating the standard deviation of the target values that reach a node as a measure of the error at that node, and calculating the expected reduction in error as a result of testing each input variable at that node. The expected error reduction, which is called standard deviation reduction (SDR) is calculated by:

$$\text{SDR} = sd(T) - \sum_i \frac{|T_i|}{|T|} sd(T_i) \tag{2-12}$$

where, T represents the set of examples that reach the splitting node, T_1, T_2,..., represents the subset of T that results from splitting the node according to the chosen input variable, sd represents standard deviation,and $|T_i|/|T|$ is the weight that represents the fraction of the examples belonging to subset T_i.

After examining all possible splits, M5 chooses the one that maximizes SDR. The splitting of the training examples is done recursively to the subsets. The splitting process terminates either when the target values of all the examples that reach a node vary only slightly, or when merely a few instances remain. This relentless division often produces overly elaborate structures that require pruning, for instance by replacing a subtree with a leaf. In the final stage, 'smoothing' is performed to compensate for the sharp discontinuities that will inevitably occur between the adjacent linear models at the leaves of the pruned tree. In smoothing, the outputs from adjacent linear equations are updated in such a way that their difference for the neighboring input vectors belonging to the different leaf models will be smaller. Details of the pruning and smoothing process can be found in Witten and Frank (2000). Figure 2-22 presents an example of model trees.

When compared to other machine learning techniques, model trees learn efficiently and can tackle tasks with very high dimensionality – up to hundreds of variables. The main advantage of model tree is the results are transparent and interpretable. Solomatine and Siek (2006) proposed two new versions of the M5 algorithm. One is the M5opt, which allows for deeper optimization of trees, and other is the M5flex that gives a modeller more possibilities to decide how the data space shold be split in the process of building regression models.

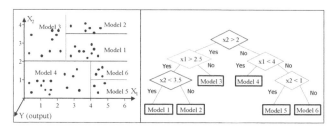

Figure 2-22. Example of model trees.

2.5.5 Locally weighted regression

Many machine learning methods including those described in Sections 2.5.3 and 2.5.4 are model-based methods. This means that they explicitly construct the model to predict or classify the output as soon as the training examples are provided. After training, the model is used for predictions and the data are generally discarded. In contrast, instance-based learning (IBL) simply stores the presented training data and postpone using it until a new instance or input has to be predicted or classified. When a new input vector is presented to the model, a subset of similar instances is retrieved from the previously stored examples and their corresponding outputs are used to predict or classify the output for the new query vector (instance). IBL methods, in fact, construct a local approximation of the modeled function that applies in the neighborhood of the new query instance (input vector) encountered. They never construct an approximation designed to perform well over the entire input space. Therefore, even a very complex target function can be described by constructing it from a collection of much less complex local approximations.

IBL algorithms have several advantages: they are simple yet robust learning algorithms, they can tolerate noise and irrelevant attributes, they can represent both probabilistic and overlapping concepts, and they naturally exploit inter-attribute relationships (Aha et al.,

1991). However, they also have disadvantages. One of them is that the computational cost of predicting or classifying new instances can be high because all computations take place on-line when the new instances have to be classified or predicted, rather than when the training examples are first encountered. The computational cost increases with the amount of training data and the dimension of the input data and has order of $|T| \times p$, where T is the training set and p is the dimension of the input vector. Furthermore, because all attributes of the examples are considered when attempting to retrieve similar training examples from the stored database, the examples that are truly most similar may well be a large distance apart if the target variable only depends on only a few of the many input variables.

The most common IBL methods used in numeric prediction are the k-Nearest Neighbour method and locally weighted regression. For a detailed description of IBL methods, the readers are referred to Aha et al. (1991). The application of these methods in rainfall-runoff modelling was reported in Solomatine et al. (2006, 2008).

Locally weighted regression (LWR) is a memory-based method for performing a regression around a point x_q of interest (often called a query point, or a new input vector) using only training data that are "local" to that point. The training examples are assigned weights according to their distance from the query instance, and regression equations are generated using the weighted data. The so-called "locally weighted regression" is deemed *local* because the function is approximated based on data near the query point. It is deemed *weighted* because the contribution of each training example is weighted by its distance from the query point. The target function f in the neighbourhood surrounding the query point x_q can be approximated using a linear function, a quadratic function, neural network, etc.

A number of distance-based weighting schemes can be used in LWR (Scott, 1992). A common choice is to compute the weight w_i of each instance x_i according to the inverse of their Euclidean distance $d(x_i, x_q)$ from the query instance x_q as given by

$$w_i = K(d(\mathbf{x}_q, \mathbf{x}_i)) = (d(\mathbf{x}_q, \mathbf{x}_i))^{-1} \tag{2-13}$$

where $K(.)$ is typically referred to as the kernel function. The Euclidean distance $d(x_i, x_q)$ is define as

$$d(\mathbf{x}_i, \mathbf{x}_q) = \sqrt{\sum_{r=1}^{p} (a_r(\mathbf{x}_i) - a_r(\mathbf{x}_q))^2} \tag{2-14}$$

where an arbitrary instance x can be described by the feature vector $<a_1(\mathbf{x}), \ldots, a_p(\mathbf{x})>$ and $a_r(\mathbf{x}_i)$ denotes the value of the rth attribute of the instance x_i.

Alternatively, instead of weighting the data directly, the model errors for each instance used in the regression equation are weighted to form the total error criterion $C(q)$ to be minimized:

$$C(q) = \sum_{i=1}^{|T|} L\left(f(\mathbf{x}_i, \beta), y_i\right) K(d(\mathbf{x}_i, \mathbf{x}_q)) \tag{2-15}$$

where $f(\mathbf{x}_i, \beta)$ is the regression model; $L(.)$ is the loss or error function (typically the sum of squared differences $(y_i - \hat{y}_i)^2$ between the target y_i and its estimated $\hat{y}_i$ values), and β is a vector of parameters or coefficient of the model to be identified. Note that the total error criterion is summed over the entire set T of the training examples. However there are other forms of the error criteria, such as the loss function over just the k-Nearest Neighbours (Mitchell, 1997). In locally weighted regression, the function f is a linear function of the form:

$$f(\mathbf{x}_i, \beta) = \beta_0 + \beta_1 a_1(\mathbf{x}_i) + \dots + \beta_p a_p(\mathbf{x}_i) \qquad (2\text{-}16)$$

Cleveland and Loader (1994) among other addressed the issue of choosing weighting (kernel) functions: the function should be at it maximum at zero distance, and the function should decay smoothly as the distance increases. Discontinuities in the weighting functions lead to discontinuities in the predictions because the training points cross the discontinuity as the query changes. Yet another possibility to improve the accuracy of LWR is to use the so-called smoothing, or bandwidth parameter, which scales the distance function by dividing it by this parameter. One way to choose the smoothing parameter is to set it to the distance to the kth nearest training instance; thus its value becomes smaller as the volume of training data increases. Generally, an appropriate smoothing parameter can be found using cross-validation.

2.5.6 Selection of input variables

Input variables for building machine learning models contain information about the complex (linear or nonlinear) relationship with the model outputs (see, e.g., Guyon and Elisseeff, 2003; Bowden et al., 2005). Generally, correlation analysis and mutual information analysis are used to determine the strength of the relationship between the input time series variables and the output time series at various lags. Correlation analysis is used to find the linear relationship between the variables and mutual information analysis is used to determine linear or non-linear dependencies. Correlation analysis computes a cross correlation between the input vector $\mathbf{x}_i$ and the output variable y. For the time series variable, it is often required to compute (i) correlation of the lagged vector of $\mathbf{x}_i$ with the output y; and (ii) the autocorrelation of the output vector y. The former measures the dependency of the output variable to the previous values of the input variables. The latter provides the information on the dependency of the output variable on its past values.

The correlation coefficient (CoC) between input vector x_t and output vector y_t, $t=1, \dots, n$ is given by:

$$CoC = \frac{\sum_{i=1}^{n}(x_i - \bar{x})(y_i - \bar{y})}{\sqrt{\sum_{i=1}^{n}(x_i - \bar{x})^2}\sqrt{\sum_{i=1}^{n}(y_i - \bar{y})^2}} \qquad (2\text{-}17)$$

where $\bar{x}$ is the mean of x. The maximum value of CoC is 1 for complete positive correlation and the minimum value of CoC is -1 for complete negative correlation. A CoC value close to zero indicates that the variables are uncorrelated.

The mutual information based on Shannon's entropy (Shannon, 1948) is measured to investigate the linear and nonlinear dependencies and lag effects (in time series data) between the variables. The mutual information is measure of the information available from one set of data having knowledge of another set of data. The average mutual information (AMI) between two variables X and Y is given by:

$$\text{AMI}=\sum_{i,j} P_{XY}(x_i,y_j)\log_2\left[\frac{P_{XY}(x_i,y_j)}{P_X(x_i)P_Y(y_j)}\right] \tag{2-18}$$

where $P_X(x)$ and $P_Y(y)$ are the marginal probability density functions of X and Y, respectively, and $P_{XY}(x,y)$ is the joint probability density functions of X and Y. If there is no dependence between X and Y, then by definition the joint probability density $P_{XY}(x,y)$ would be equal to the product of the marginal densities ($P_X(x)\,P_Y(y)$). In this case, AMI would be zero (the ratio of the joint and marginal densities in Equation (2-18) being one, giving the logarithm a value of zero). A high AMI value indicates a high level of dependence between two variables. The key to accurately estimating of the AMI lie in the accurately estimating the marginal and joint probability densities in Equation (2-18) from a finite set of examples.

2.5.7 Data-driven rainfall-runoff model of the Bagmati catchment

Nest , we construct a rainfall-runoff model of the Bagmati catchment. *CoC* and AMI are used to select the most important input variables. Figure 2-23 shows the correlation coefficient and the AMI of RE_l as well as its lagged variables. It is observed that the correlation coefficient is highest at 1 hour and the optimal lag time (time at which the correlation coefficient and/or AMI are the maximum) is also 1 hour. At this optimal lag time, the variable RE_l provides a maximum amount of information about the output. Additionally, the correlation coefficient and AMI of the observed discharge are analyzed. The results show that both the immediate and the recent discharges (with the lag of 0, 1, 2 h) have a very high correlation.

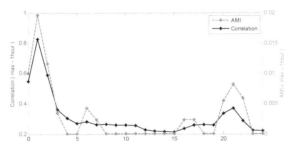

Figure 2-23. Simulated linear correlation between rainfall and runoff

Based on the above analysis, we consider several input data structures for the machine learning models. With the maximum dependence of 0 hours and 1 hours lag time, two machine learning models are used (namely ANNs and MT)). The same data sets used for calibration and verification of the HBV model are used for training and verification respectively. However, for proper training of the ANN models the calibration data set is segmented into two subsets: 15% of data sets for cross validation (CV) and 85% for training (see Figure 2-9). The CV data set is used to identify the best structure for the ANN models.

In ANN, a multilayer perceptron network with one hidden layer is used; optimization is performed by the Levenberg-Marquardt algorithm. The hyperbolic tangent function is used for the hidden layer with a linear transfer function at the output layer. The maximum number of epoch is fixed to 1000. A trial and error method is adopted to detect the optimal number of neurons in the hidden layer, testing various numbers of neurons from 2 to 10 as well as different inputs. We observed that a layer with 10 neurons yields the lowest error on the CV data set and the best input set has 5 variables $RE_{t-0}, RE_{t-1}, RE_{t-2}, Q_{t-1}, Q_{t-2}$.

Figure 2-25 shows a fragment of the validation period; on 20 Aug 1990, one of the peak of hydrograph peaks was improved by ANN model in camparison to both the HBV and MT models.

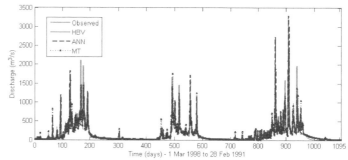

Figure 2-24. The hydrographs of observed discharge, and simulated discharge by HBV, ANN and MT in verification

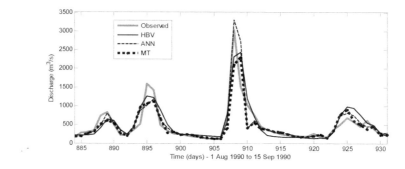

Figure 2-25. The fragment of discharge simulated by HBV, ANN and MT in verification

Table 2-4. Performances of HBV, ANN and MT models for the Bagmant catchment.

Models	Calibration		Verification	
	NSE	RMSE	NSE	RMSE
HBV	0.87	102.63	0.83	111.56
ANN	0.95 tr/ 0.89 cv*	62.08 tr/ 93.21 cv	0.81	113.46
MT	0.91	79.57	0.81	110.50

- *tr indicates training ; cv indicates cross-validation

The MT experiment is carried out with different numbers of pruning factor in order to controls for the complexity of the generated models. We report the results of an MT that has a moderate level of complexity. Note that the CV data set has not been used in MT, rather the

MT uses the full calibration data set to build the models. We observed that the value of pruning factor 2 yields 8 local models with best performance. The results are shown in Table 2-4.

In Table 2-4, the performances of ANN and MT have high value of *NSE* and *RMSE* during the calibration period compared with HBV, however these values were yielded low in verification period. The *NSE* values were similar during verification period.

2.5.8 Data-driven rainfall-runoff model of the Leaf catchment

We have also constructed a rainfall-runoff model for the Leaf catchment. Based on the correlation and AMI analyses, we considered several input data structures, Ultimately, RE_{t-7}, RE_{t-8}, RE_{t-9}, Q_{t-1}, ΔQ_{t-1} yielded better results for predicting the streamflows. Figure 2-26 shows a comparison of the HBV, ANN and MT. None of all three models fails to cover the peak (Figure 2-26). The MT was marginally better than the other two models in terms of performance measures. Table 2-5 shows the *NSE* and *RMSE* values predicted by the two machine learning models.

Table 2-5. Performances of HBV, ANN and MT models for the Leaf catchment

Models	Calibration		Verification	
	NSE	RMSE	NSE	RMSE
HBV	0.87	17.56	0.90	26.76
ANN	0.95 tr/ 0.92 cv*	12.08 tr/ 14.21 cv	0.89	27.46
MT	0.90	15.57	0.91	25.50

- *tr indicates training ; cv indicates cross-validation

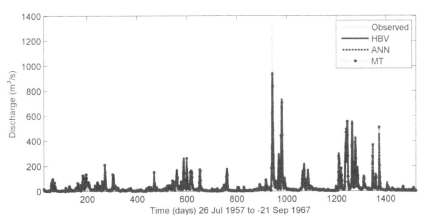

Figure 2-26. The hydrographs of observed discharge and simulated discharge by HBV, ANN and MT models in verification

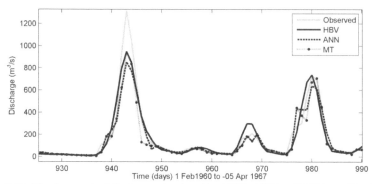

Figure 2-27. The fragment of discharge simulated by HBV, ANN and MT in verification.

2.6 Summary

Brief descriptions of hydrological models, both conceptual and data driven, and their calibration have been discussed in this chapter. The HBV conceptual hydrological model has been setup for the Alzette, Bagmati, Brue, Leaf, and Nzoia catchments and SWAT semi-distributed hydrological model has been set up for Nzioa catchment (Kenya). Data driven models were set up for the Bagmati and Leaf catchments. Additionally quantification of the model output errors has been given as well.

Chapter 3
Committees of hydrological models

This chapter contributes to refining the committee (multi-model) approach to hydrological modelling. It deals with a dynamic combination of hydrological models by the so-called "fuzzy committee" method. First, it presents the committee models by using several weighting schemes used in objective functions for calibration of specialized models, as well as different membership functions to combine models. Secondly, it presents a combination of specialized models where their weights depend on the hydrological model state variables (soil moisture, base flow, etc.), inputs (precipitation and evapotranspiration), and outputs (simulated streamflows), which are different for each time step, depending on the current value of flow. This chapter also presents the performance of various committee models and their comparisons.[1]

3.1   Introduction

The conceptual hydrological model is composed by fluxes and storages representing relevant complex hydrological processes of the catchment. This model is used to predict the behaviour of time-varying streamflows, but the strength of the predictions depends on the presumed model structure, described parameters, and quality of data used. In usual practice, modellers often assume that the data being fed into the hydrological model's (conceptually lumped) overall structure are correct, and the model prediction is deliberately presented based on measurement data using the degree of knowledge and experience by discovering the optimum best parameter set. Typically, this approach focuses on a single model using the single best set of parameters. However, the model produced by one best set of parameters might not equally well describe the characteristic of the hydrological processes for all ranges of flow. Furthermore, different models have strength in predictions of different processes (aspects of hydrological responses), but a single model in isolation has difficulty handling all the processes. (Note that the widely used notion of a model "ensemble" is in a way a particular case of a committee – in an ensemble each model is responsible for the whole modelled process but they are differently parameterised or use different initial conditions.)

Multi-model averaging is one way to improve the performances of model prediction and, it has been receiving a surge of attention related to deriving predictive model output. This approach involves a combination of several individual models into a new single model, where each individual model's strength is presented in such way that optimal prediction can be obtained, and the weaknesses of each individual model are compensated for by each other. Multi-model approaches are not new in hydrological modelling – examples are the early works of Keefer and McQuivey (1974), Todini and Wallis (1977), Bruen (1985) and Becker

---

[1] Kayastha, N., and Solomatine, D. P. (2014) Committees of specialised conceptual hydrological models: comparative study, *11th International Conference on Hydroinformatics*, New York USA

Kayastha, N., Ye, J., Fenicia ,F., Kuzmin,V., and Solomatine, D. P. (2013). Fuzzy committees of specialized rainfall-runoff models: further enhancements and tests, *Hydrol. Earth Syst. Sci.*, 17, 4441-4451.

and Kundzewicz (1987) (the latter authors built piecewise linear models instead of the overall linear hydrological model). Cavadias and Morin (1986) aggregated several watershed models, which were considered by WMO (1986) for intercomparison of their model performances. Juemoe et al. (1987) combined a conceptual model and a statistical model, which is known as a synthesized constrained linear systems model. This model was developed by a combination of the Xinanjiang model (Zhao, 1977) and the constrained linear system model (Todini and Wallis, 1977). McLeod et al. (1987) combined three models, namely a transfer function noise model, a periodic autoregressive model, and a conceptual model for flow forecast. Since then, various authors have explored different approaches to identify diverse hydrological regimes and ways of combining specialized models that are both process-based and data-driven (Table 3-1 and Table 3-2). Many authors introduced combinations of multiple hydrological models and concluded that a blend of models is superior to the predictions of unbiased single models. A number of methods for combining models are summarized in Table 3-1 and Table 3-2.

Table 3-1. The model combination methods proposed in earlier works

Authors	Combination methods	Combined models
Shamseldin et al. (1997)	Simple average method (SAM), weighted average method (WAM), and neural networks (NNs) method	SLM, LPM, LVGFM, CLS-Ts, and SMAR
Shamseldin and O'Connor (1999)	Linear Transfer Function (LTF) and WAM	LPM, LVGFM, and SMAR
See and Openshaw (1999)	Rule-based fuzzy logic	ANN model (Self-Organizing Maps (SOM), and multi-layer perceptron neural (MLP)
See and Openshaw (2000)	SAM, Bayesian approach (BA), and Fuzzy Logic (FL) models	Hybrid NN, a simple rule-based FL model, an ARMA model, and naive predictions model
Xiong et al. (2001)	First-order Takagi–Sugeno fuzzy system	SLM, LPM, LVGFM, CLS-Ts, and SMAR
Abrahart and See (2002)	Arithmetic-average	NNs model of low, medium, high, very high, and peak flows, FL model, TOPMODEL, ARMA model
Coulibaly et al. (2005)	Weighted-average	Nearest-neighbor model, conceptual model, and ANN model
Chen and Adams (2006)	ANNs with back-propagation algorithm	HBV-S sub-basin models
Kim et al. (2006)	SAM, constant coefficient regression, switching regression, sum of squared error, and ANN	Tank model and NNs model
Solomatine (2006)	Fuzzy committee	High- and low-flow model by Model tree

Ajami et al. (2006)	SAM, the multimodal superensemble (MMSE), modified multimodel superensemble (M3SE), and WAM.	DMIP models (Smith et al., 2004)
Oudin et al. (2006)	Time-varying combinations method	High- and low-flows from GR4J and TOPMODEL
Duan et al. (2007)	Bayesian model averaging (BMA)	SAC-SMA, Simple Water Balance (SWB) model, and HYMOD
Shamseldin et al. (2007)	Simple neural network (SNN), radial basis function neural network (RBFNN) and multi-layer perceptron neural network (MLPNN)	SMAR , PDISC, LPM and LVGFM
Fenicia et al. (2007) Cullmann et al. (2008)	Fuzzy committee Weighted averaging based on sigmoid weight functions	High and low-flow by HBV Event class by WaSiM-ETH
Nasr and Bruen (2008)	Neuro-fuzzy model with Takagi–Sugeno fuzzy approach	Simple linear model (SLM) and the soil moisture and accounting routing (SMAR)
Devineni et al. (2008)	Rank probability score skill	Statistical models (climatological ensembles)
Viney et al. (2009)	SMA, MLR, BMA, WAM, conditional ensembles based on flow stage (i.e., rising or falling), conditional ensembles based on flow level (i.e., high and low flows).	DHSVM ,MIKE-SHE,TOPLATS, WASIM-ETH, SWAT, PRMS, SLURP, HBV, LASCAM and IHACRES
Jeong and Kim (2009)	SAM, WAM, Regression and ANN	SSARR, TANK, abcd,GR2M, and ENN
Velazquez et al. (2010)	SAM, GA, Continue Rank Probability Score (CRPS)	GR4J, PDM, MORD, TOPM, SACR, SMAR, NAM, TANK, HBV, CREC, WAGE, IHAC 6, GARD, SIMH, MOHY, CEQU, HYM
Hostache et al. (2011)	Fuzzy committee	High- and low-flow by FLEX

The data-driven models (DDM) are widely used in rainfall runoff modelling. These models have capabilities of learning from data without requiring prior knowledge of the model component process. Artificial Neural Networks (ANN) are very popular for the application of DDM, to build models and/or sub-models. Refer to Maier et al. (2010) for a comprehensive review, where numerous studies have demonstrated the modelling of rainfall runoff processes. In the ANN model, the training set is split into a number of subsets, and separate models are trained on these subsets (Corzo and Solomatine 2007b).

The problem of complex modelling can be solved by splitting it into several simple tasks and building a simple model for each of them. In this respect, the input (state) space can be divided into a number of regions, for each of which a separate specialized model is built (Figure 1-1). These specialized models are also called local, or expert models, and form a modular model (MM). In machine learning, MMs are attributed to a class of committee machines. There are several available ways to classify committee machines; Solomatine and Siek (2006) proposed the method of learning MM using the input data set split (hard split) into sub-sets, which results to MMs, and then combining the soft splits of booting schemes (e.g., AdaBoost.RT proposed by Shrestha and Solomatine, 2004).

Solomatine (2005) classified committee machine into four categories (i) hard splitting where individual experts are trained on particular subspaces of data independently and use only one of them to predict the output for a new input vector, e.g. model trees, regression trees (Breiman et al., 1984); (ii) hard splitting – soft combination where outputs are combined by 'soft' weighting scheme; (iii) statistically-driven soft splitting, used in mixtures of experts (Jacobs et al., 1991) and boosting schemes (Freund and Schapire, 1996; Shrestha and Solomatine, 2006); and (iv) no-split option leading to ensembles; the models are trained on the whole training set and their outputs are combined using a weighting scheme where the model weights typically depend on model accuracy, e.g. bagging (Breiman, 1996). The schema of MM (local models) and their combination are shown in Figure 1-1 in Chapter 1.

Inspite of the limitations of DDM (they do not provide any information on internal hydrological processes during predictions of runoff), they can be very useful for river flow forecasting (e.g., Nayak et al., 2005; Chapter 2). Several studies (refer to Table 3-2) focussed on DDM for building individual models (local models or sub-models) and combining them in order to provide more accurate predictions.

Table 3-2. Combined data-driven models proposed in earlier works

Authors	Methods used	Combined models
See and Openshaw (1999)	Rule-based fuzzy logic	ANN (SOM and MLP)
Abrahart and See (2002)	Fuzzy logic	Low, medium, high, very high, and peak discharges by NNs
Solomatine and Xue (2004)	NNs and M5 model trees for different flow regimes	ANN and M5
Anctil and Tape (2004)	Bayesian regularization, bagging and stacking	MLP and LM
Nilssom et al. (2005)	Snow accumulation and soil moisture calculated by conceptual model as input to NNs.	HBV and NNs
Solomatine and Siek (2006)	Fuzzy committees	M5 local models (by hierarchical splitting of hydrological data)
Jain and Srinivasulu (2006)	NNs for various flows regimes	MLP and SOP model of decomposed flows
Corzo and Solomatine (2007a)	Separate NNs for various flow regimes	ANN model of various flow regimes, base flow and direct flow.
Corzo and Solomatine (2007b)	Separate NNs for various flow regimes	ANN model of base flow and excess flow.
Toth (2009)	Clustering based on SOM	ANNs for each data subset
Boucher et al. (2010)	Mean CRPS.	Ensemble of MLP

One way to improve predictions of the single hydrological model is to model different processes separately, with each representing a particular process, and merging them to produce a combined model. In this type of modelling paradigm, every model can be differently and specifically oriented to a particular process, or the same model structure can be used, but calibrated differently for different regimes of the same process. These specifically-oriented models can then be combined with time-varying weights (dynamic weighted averaging). Fenicia et al. (2007) presented two weighted performance measures for selected characteristics of hydrographs (high and low flow), and then weighted them together using a fuzzy combining scheme proposed by Solomatine (2006). Their work was extended by Kayastha et al. (2013), who considered the suitability of different weighting functions for objective functions and different classes of membership functions used to combine the local models and compared them with global optimal models. This approach was also used by Hostache et al. (2011) to calibrate two separate low-flow and high-flow conditions using the last 24 hours and 120 hours ahead, using forecasted rainfall and observed temperature as inputs. The forecasted discharge for 120 hours is obtained by using a combining function. They suggested that the combining local model approach could improve the performance of hydrological model forecasting.

The different parameterizations within the same model structure that consyitutes one model could be better represented to either high-flow regimes and the other to low-flow regimes, but the same parameterization could not successfully address in both cases. In this respect, the flow series can be split into periods covering different parts of catchment responses (e.g., wet periods and dry periods, excess flows and base flows, high flows and low flows, rising and falling limbs) and then their performance measure can be calculated separately. Willems (2009) demonstrated the application of the multi-criteria model evaluation procedure in a river flow series in which the flow series are separated into: (i) sub-flows, using the filter technique; (ii) independent quick and slow-flow periods; and (ii) independent peak and low-flow values. This type of modelling provides knowledge on certain important local properties, which reveal useful insights into underlying problems. However, this technique is time-consuming and requires specific domain knowledge on flow characteristics. Corzo and Solomatine (2007a, b) used several types of filters to separate flows into base and excess flow and to build separate data-driven forecasting models for each of these types of flow. A genetic algorithm optimizes the weights for the combination scheme.

Oudin et al. (2006) presented the model combination approach, which is based on time-varying weights, and these are estimated as: (i) simple average (fixed weights); (ii) based on extraterrestrial radiation (climatic factor); and (iii) based on the state of variables of the rainfall runoff model. These weights are used to combine two differently-parameterized models of the same structure. They reported that the combined models could produce a reasonable outcome compared with a single model. In spite of this, the combined model, based on weights derived from extraterrestrial radiation, did not perform well because these weights do not consider daily hydro-climatic conditions (This is considered only on inter-annual index.). Their results showed that a combination of models based on time-varying weights (from state variable -soil moisture accounting store) performed better than fixed weights because the nature of the error differs in time, and these weights dynamically adjust the error between candidate models. One should note that climatic factors (temperature and evapotranspiration) typically play a minor role in describing the nature of streamflows, but were strongly influenced by infiltration, storage capacity, and intensity and duration of rainfall (Jain and Srinivasulu, 2006).

Marshall et al. (2007) used the so-called Hierarchical Mixture of Experts approach known from machine learning in which a special "gating function" makes a switch between specialized (expert) models at every time step. The variable dictating the probability of such a switch is the modelled subsurface catchment storage. The justification for this is that the mechanisms for producing flow in a wet period, with full storage, are different from those in a dry period when the catchment is empty (This reasoning is similar to that in Oudin et al., 2006). Note that in this approach, there was no combination of models' outputs at each time step, as the investigators were using only one of them.

Nasr and Bruen (2008) used a neuro-fuzzy model to combine a number of sub-models to represent the temporal and spatial variation in catchment response. The different sub-models are built under both flood conditions and drier conditions. Each input value (rainfall) is assigned to a particular sub-set, and memberships (weights) are estimated. Thus, the weights are identified based on the rainfall (model input). Cullmann et al. (2008) presents a process-based calibration where the hydrological model is calibrated based on the classification of specific events (for extreme and normal floods). The event-specific models are combined using the sigmoid weighting scheme. The objective of their method is to predict flood peaks by integrating two different classes of flood events. However, low flows were not considered.

This chapter presents various approaches of committee modelling to improve the hydrological model prediction. The first part of this chapter explores the fuzzy committee method that was initially proposed by Fenicia et al. (2007). In this approach, the weights assigned to each specialized model's output are based on optimally-designed fuzzy membership functions, and they are different at every time step. Furthermore, this study reports the results of various committee models that used several weighting schemes in objective functions for the calibration of specialized models, as well as different membership functions to combine models (Kayastha et al., 2013). Fuzzy committee models are tested for Alzette, Bagmati, and Brue catchments, while these models used two approaches of optimization for the calibration of models: (i) multi-objective optimization by the Non-dominated Sorted Genetic Algorithms (NSGA II) by Deb et al. (2002) used to find Pareto optimal solutions of specialized models; (ii) Single objective optimization - Genetic Algorithm (GA) (Goldberg (1989) and Adaptive Cluster Covering Algorithm (ACCO) by Solomatine (1999) are used to calibrate the single specialized models and single global models. The second part of this chapter explores different committee models that were built by a dynamic combination of specialized models using weights: (i) based on hydrological states (e.g., Soil moisture accounting store (Oudin et al., 2006) and other states (e.g., quick and slow flows); (ii) based on model inputs; and (iii) based on model outputs (simulated flows). These models were tested for the Brue, Bagmati, and Leaf catchments.

3.2   Specialized hydrological models

As it has been mentioned, the predictions of hydrological models are often set based on a single model, and their evaluation is done by single aggregate measure criteria (e.g., *RMSE*). However, these models typically not fully capable to capture all the characteristics of streamflows. The reason for this is that the characteristic of streamflows always vary by several orders of magnitude and their variance of error is dependent on the flow value.

Hydrological models are composed of complex systems (processes) that are comprised of multiple components (simpler sub-processes). Therefore, instead of one model, several sub-models can be composed that represent sub-processes separately, where each model will then represent a specific process and time regime (hydro-meteorological situation). The sub-models are also called "specialized models". The idea of building specialized models is to accurately reproduce the catchment hydrological responses (streamflows). This is accomplished by assembling an individual single model that is instantiated in the same model structure, specialized on distinctive regimes (high flows and low flows) of system behaviour, and their performances are evaluated by different objective functions. At this point, the weighted objective functions can be used to evaluate performance measures, where these objective functions stress the model errors with respect to each flow simulation. Alternatively, individual models can be composed by transformation of the objective function, emphasizing model parameterization for efficient regimes of flow domains. For example, the model could be better adapted to low flows by using the transformation of the objective function, which helps, to a certain extent, in avoiding imperfect models (Oudin et al. 2006).

In this study we aim at refining the fuzzy committee approach (Solomatine 2006; Fenicia et al. 2007), and the main results are presented in publication by Kayastha et al. (2013). This approach is briefly outlined below. The high flows and low flows are considered as distinctive regimes, or states of system behaviour. The main aim is to accurately reproduce the system response during both regimes, for each of which separate models are built. In order to evaluate the performance of the single hydrological model in both conditions, the two weighted objective functions are used, where one emphasises the model error with respect to low-flow simulation, and the other is focuses on the model error with respect to high flows.

The two objective functions are defined as follows:

$$RMSE_{1LF} = \sqrt{\frac{1}{n}\left(\sum_{i=1}^{n}(Q_{s,i} - Q_{o,i})^2 . W_{LF,i}\right)}$$
(3-1)

$$RMSE_{HF} = \sqrt{\frac{1}{n}\left(\sum_{i=1}^{n}(Q_{s,i} - Q_{o,i})^2 . W_{HF,i}\right)}$$
(3-2)

where n is the total number of time steps, $Q_{s,i}$ is simulated flow for the time step i, $Q_{o,i}$ is observed flow for time step i. The two weighting functions, W_{LF} and W_{HF}, allow for placing the stronger weight on the low or on the high portions of the hydrograph. As a result, $RMSE_{LF}$ places higher weight on low flow errors and lower weight on high flow errors than $RMSE_{HF}$. Please note that values $RMSE_{LF}$ and $RMSE_{HF}$ cannot be compared with each other and with the values of $RMSE$ because of differences in weighting; this is important when viewing the resulting plots (Figure 3-4 and Figure 3-5). The types of the weighting functions (schemes) together with their parameters will be referred to further as $WStype$. Corresponding equations and figures are given below.

Table 3-3. Types of the weighting functions

WStype	$W_{LF,i}$	$W_{HF,i}$	Eq. No	Figures
I	$= (l)^N$	$= (h)^N$	(3-3)	Figure 1a
II	$= \begin{cases} 0, & if \quad l > \alpha \\ (1-l*(1/2-\alpha))^N, & if \quad l \leq \alpha \end{cases}$	$= \begin{cases} 1, & if \quad h > \alpha \\ (h/\alpha)^N, & if \quad h \leq \alpha \end{cases}$	(3-4)	Figure 1b
III	$= \begin{cases} 0, & if \quad l > \alpha \\ (1-l*(1/2-\alpha))^N, & if \quad l \leq \alpha \end{cases}$	$= \begin{cases} 1, & if \quad h > \alpha \\ 0, & if \quad h \leq \alpha \end{cases}$	(3-5)	Figure 1c
IV	$= \begin{cases} 0, & if \quad l > \alpha \\ (1-l*(1/2-\alpha)), & if \quad l \leq \alpha \end{cases}$	$= \begin{cases} 1, & if \quad h > \alpha \\ 0, & if \quad h \leq \alpha \end{cases}$	(3-6)	Figure 1d

In above equations (Table 3-3), the variables l and h are calculated as:

$$l = \frac{Q_{o,\max} - Q_{o,i}}{Q_{o,\max}} \qquad h = \frac{Q_{o,i}}{Q_{o,\max}} \qquad (3\text{-}7)$$

where $Q_{o,max}$ is the maximum observed flow, N is the power value (for the experiment in this study, we considered only 1, 2 or 3), and α is a threshold for selecting weights of flows (It was chosen to be 0.75). Note that both N and α can also be subjected to optimization, but, in this study, that was not done. By computing both objective functions over the full range of discharges, both functions constrain the model to fit the entire hydrograph for WStype I, where parameter α is not used. However, for WStype with parameter α, W_{LF} excludes high flows from the computation of the objective function if the condition is $l > \alpha$. In the same way, W_{HF} excludes low flows if $h \leq \alpha$ for WStype III and IV.

As mentioned above, measured criteria based on the mean square error place greater emphasis on high flow simulation and may not be suited for low flows. The objective function can be selected for the choice of a target variable in calculating criteria. Oudin et al. (2006) proposed an objective function based on the logarithms for transformations on low-flow values. This criteria is used to calibrate low flows and illustrated in the following as logarithmic root mean square:

$$RMSE_{\ln} = \sqrt{\frac{1}{n}\left(\sum_{i=1}^{n}(\ln Q_{s,i} - \ln Q_{o,i})^2\right)} \qquad (3\text{-}8)$$

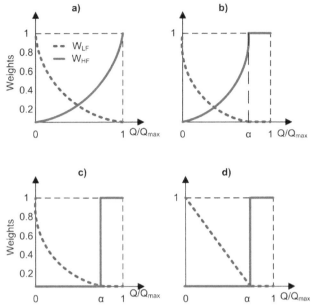

Figure 3-1. (a) *WStype* I, -weighting scheme for objective functions studied in Fenicia et al. (2007); (b) *WStype* II, (c) *WStype* III; and (d) *WStype* IV. Note: These three additional weighting schemes were attempted in the latest experiments.

3.3 Committees of specialized models

It is possible to classify various types of committee models into the four groups, depending on the weighting scheme.

3.3.1 Fuzzy committee models

The specialized models are built under the conditions of different regimes of catchment hydrological responses and are combined using an appropriate combining scheme. Although this combining scheme can be straightforward (Jain and Srinivasulu 2006), implying a switch between different models at different time steps, it can also involve various methods of model averaging (refer to Table 3-1). However, the issue is how to handle the compatibility at the boundaries between the two different specialized models. One possible way is to use a soft weighting scheme that switches to a smooth transition between the boundaries, where the contribution of each specialized model is based on using a fuzzy membership function – the so-called "fuzzy committee" described by Solomatine (2006). Xiong et al. (2001) also employed the fuzzy combination (Takagi-Sugeno); of models but their method was used to integrate the ensemble models (handling the whole range of the flow conditions), whereas the fuzzy committee integrates models were specifically developed for different flow conditions.

The shape of the function initially used was trapezoidal (later another shape, see Figure 3-2 b) that parameterized by the two transitional parameters [γ, δ] (see Figure 3-1a). The membership function (weight) of the low-flow model was assigned 1 when the relative flow was below the parameter γ, it started to decrease in the proximity of the region boundary when the relative flow was between γ and δ; and it decreased to zero beyond the boundary when the relative flow was above δ. The membership function of the high-flow model follows the opposite logic. These membership functions are described by Equation 3-10 and 3-11. The outputs of models are multiplied by the weights that depend on the value of flow and then normalized (Equation 3-9). The overall committee model is defined as follows:

The committee model defines as follow:

$$Q_{c,i} = (m_{LF} \cdot Q_{LF,i} + m_{HF} \cdot Q_{HF,i})/(m_{LF} + m_{HF}) \tag{3-9}$$

$$m_{LF} = \begin{cases} 1, & if \quad h < \gamma \\ 1-(h-\gamma)/(\delta-\gamma)^N, & if \gamma \leq h < \delta \\ 0, & if \quad h \geq \delta \end{cases} \tag{3-10}$$

$$m_{HF} = \begin{cases} 0, & if \quad h < \gamma \\ (h-\gamma)/(\delta-\gamma)^{1/N}, & if \gamma \leq h < \delta \\ 1, & if \quad h \geq \delta \end{cases} \tag{3-11}$$

where m_{LF} and m_{HF} are membership functions for the two individual models, $Q_{LF,i}$ and $Q_{HF,i}$ simulate high and low flows respectively for the time step i; γ and δ are the thresholds for high and for low flows respectively, and N is the parameter determining the steepness of the function used. The value $N=1$ is given for Type A and value of 2 or greater are assigned for Type B. The first two optimal specialized models, that is, one for the low-flow ($Q_{HF,i}$), and one for the high-flow ($Q_{HF,i}$) are sought using optimization algorithms. Next two membership function parameters δ and γ are introduced to combine specialized models, which ruled the transition between the specialized models. The committee model Q_c is calculated by combining sets of δ and γ, which are selected within given intervals, and the performance measure is calculated by *RMSE* and *NSE*.

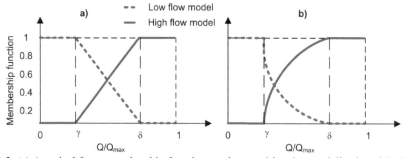

Figure 3-2. (a) A typical fuzzy membership function used to combine the specialized models (Type A); (b) A class of membership functions for high and low flow models tested in the new experiments (Type B).

3.3.2   States-based committee models

Model averaging approaches use various types of weights to combine individual model outputs, which improve simulation and eliminate unrealistic discontinuities in the simulated system behaviour. The "states-based" committee models are composed by weights that depend on the internal model variables and are used to combine specialized models, which are built under the conditions of high-low and low-flow regimes. Oudin et al. (2006) presented the dynamic weights necessary to combine two models. These weights were computed from the rate of the soil moisture accounting (SMA) store of the rainfall runoff models, which represent the average of the water content (between 0 and 1) of the two SMA stores from the models calibrated on objective function *RMSE* and objective function *RMSE*$_{ln}$. In this case, when the moisture rate is close to 1, the combined streamflow tends to be equal to the streamflow obtained with the objective function *RMSE*, and when the moisture rate is close to 0, the combined streamflow tends to be equal to the streamflow obtained with the objective function *RMSE*$_{ln}$. In addition, the cubic function was used for weighting scheme (index) to increase the influence of the variations of these weights because the SMA store is rarely completely full or empty, and these indexes vary slowly over time. States-based committee models are also composed by the weights that are acquired not only from the SMA store value but also from other internal model variables (e.g., upper zone, lower zone). Figure 3-3 presents a graphical representation of the different internal states of the hydrological model. This study uses two different weighting schemes (similar shapes of weighting scheme Figure 3-1a) for specialized models. The weights for combining two specialized models are calculated by internal variables of the two specialized models, where model 1 is obtained by observed high flows, and are closer to simulations obtained with objective function *RMSE*$_{LF}$. Model 2 is obtained by observed low flows that are closer to simulations with objective function *RMSE*$_{HF}$.

The combination models obtained with weights from internal variables are called "states-based committee model" and expressed as follows:

$$Q_{c,i} = (l_{sim} \cdot Q_{LF,i} + h_{sim} \cdot Q_{HF,i}) \qquad (3\text{-}12)$$

$$l_{sim} = \frac{Q_{st,\max} - Q_{st,i}}{Q_{st,\max}} \qquad h_{sim} = \frac{Q_{st,i}}{Q_{st,\max}}; \qquad (3\text{-}13)$$

where $Q_{LF,i}$ *and* $Q_{HF,i}$ are simulated high and low flows for time step i, Q_{st} is an internal variable of HBV models, and the two weighting functions l_{sim} and h_{sim} allow for placing the stronger weight on the low or on the high flows.

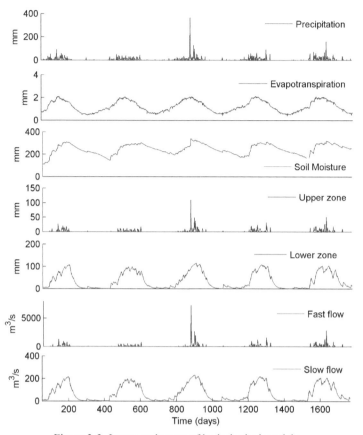

Figure 3-3. Inputs and states of hydrological models

3.3.3   Inputs-based committee models

Time-varying weights based on hydrological inputs can be also used to combine specialized models (for instance, precipitation and evapotranspiration). These weights depend on the climatic environment of the catchment. Bruen (1985) implemented the model averaging approach based on the inputs to the rainfall runoff model, which was effectively divided into a number of separate series by a threshold of fixed values, subsequently the separate sub-models were constructed. The output of the overall model was obtained by the sum of the outputs from each of the sub-models applied to the corresponding separated inputs, and these inputs corresponded to the different levels of rainfall intensity (low to high rainfall). Nasr and Buren (2008) extended this approach by estimating different levels of memberships of input to all sub-sets using the fuzzy logic principle. The degrees of memberships were taken as the weights that were assigned to the outputs from the models corresponding to each of the input sub-sets. However, in this thesis, to combine specialized models the inputs-based committee models are built without splitting (sub-sets) the input series. These weights were calculated (as in Oudin et al., 2006) from inputs as a rescaled value between 0 and 1, and then high weights were used when the model has to be more accurate for high flows, and therefore should be close to 1. In contrast, for low flows, the weight should be close to 0. The combination models are expressed as follows:

$$Q_{c,i} = (I_{low} \cdot Q_{LF,i} + I_{hig} \cdot Q_{HF,i}) \tag{3-14}$$

$$I_{low} = \frac{P_{max} - P_i}{P_{max}} \qquad I_{hig} = \frac{P_i}{P_{max}}; \tag{3-15}$$

where I_{low} and I_{hig} are weighting functions based on hydrological inputs for the two specialized models, $Q_{LF,i}$ and $Q_{HF,i}$ are simulated high and low flows for the time step i; P_{max} is the maximum value of precipitation over selected time series and P_i is precipitation for the time step i.

3.3.4   Outputs-based committee models

Outputs-based committee models are assembled by weights that are calculated based on the simulated outputs of hydrological models (simulated streamflows) to combine specialized models. The weights are calculated similar to inputs-based committee models; however, instead of inputs, the weights are calculated from the simulated streamflows of each specialized model. The combination models are expressed as follows:

$$Q_{c,i} = (Q_{Lsim} \cdot Q_{LF,i} + Q_{Hsim} \cdot Q_{HF,i}) \tag{3-16}$$

$$Q_{Lsim} = \frac{Q_{s,max} - Q_{s,i}}{Q_{s,max}} \qquad Q_{Hsim} = \frac{Q_{s,i}}{Q_{s,max}}; \tag{3-17}$$

where Q_{Lsim} and Q_{Hsim} are two weighting functions based on hydrological model outputs. $Q_{s,max}$ is the maximum simulated flow over selected time series and $Q_{s,i}$ are simulated flows for the time step i.

3.4   Performance measures

The resulting model is subsequently verified (tested) on the verification (test) data set, and compared with the single hydrological model (which is optimized by a single-objective optimization algorithm) based on *RMSE* and *NSE* (Nash-Sutcliffe efficiency (Nash and Sutcliffe, 1970) as an objective function. *RMSE* is a measure of the difference between simulated by a model and the observed that is being modelled. This helps to combine them into a single measure of predictive ability. *NSE* is a similar measure widely used in hydrology, and is calculated as 1 minus the absolute squared difference between the simulated discharges from the committee model and observed discharges normalized by the variance of the observed discharges. The value of *NSE* is in the range of [-∞, 1] and a value of 1 indicates a perfect fit of the model.

The equations of *RMSE* and *NSE* are given below:

$$RMSE = \sqrt{\frac{1}{n}\left(\sum_{i=1}^{n}\left(Q_{s,i} - Q_{o,i}\right)^2\right)} \tag{3-18}$$

$$NSE = \sqrt{1 - \frac{\sum_{i=1}^{n}\left(Q_{o,i} - Q_{s,i}\right)^2}{\sum_{i=1}^{n}\left(Q_{s,i} - \bar{Q}_{o,i}\right)^2}} \tag{3-19}$$

where $Q_{o,i}$ expresses the observed discharges for the time step i, $Q_{s,i}$ indicates the simulated discharges (single optimal or committee models) for the time step i and n is the number of observations.

3.5   Models setup for Alzette, Bagmati and Leaf catchments

Initially, the three experimental catchments, namely Alzette, Bagmati, and Leaf, were selected for testing the fuzzy committee models. In the second experiment, we tested states, inputs- and outputs-based committee models, and compared those with each other for three catchments, namely Bagmati, Leaf, and Brue. The summary statistics and records of data for calibration and verification of these catchments are presented in Chapter 1, Section 1.4.

Table 3-4. The range of model parameters

Para- meters	Units	Descriptions	Ranges used in calibration (optimization)			
			Alzette	Leaf	Bagmati	Brue
FC	(mm)	Maximum soil moisture content	100 - 450	100 - 400	50-500	100 - 300
LP	(-)	Limit for potential evapotranspiration	0.3 - 1	0.1 - 1	0.3 - 1	05 - 0.99
ALFA	(-)	Response box parameter	0.1 - 1	0 - 2	0 - 4	0 - 4
BETA	(-)	Exponential parameter in soil routine	0.1 - 2	1.0 - 4	1.0 - 6	0.9 - 2
K	(mm/h)	Recession coefficient for upper tank	0.005 - 0.5	0.05 - 0.5	0.05 - 0.5	0.0005 - 0.1
K4	(mm/h)	Recession coefficient for lower tank	0.001 - 0.1	0.01 - 0.3	0.01 - 0.3	0.0001 - 0.005
PERC	(mm/h)	Percolation from upper to lower response box	0.01 - 1	0 - 5	0 - 8	0.01 - 0.09
CFLUX	(mm/h)	Maximum value of capillary flow	0 - 0.05	0 - 1	0 - 1	0 - 0.05
MAXBAS	(h)	Transfer function parameter	8 - 15	2 - 6	1 - 3	8 15

Note: The unit 'd-day' is used for Leaf and Bagmati catchments instead of 'h-hour.'

The version of the HBV model (see Chapter 2, Section 2.2 for a description) is setup for this study. The model is calibrated using the global optimization routine named the adaptive cluster covering algorithm, ACCO (Solomatine 1999) to find the best set of parameters. The investigated sets of parameters from different models are given in Table 3-5 and Table 3-8

Table 3-5. Sets of model parameters identified by different optimization algorithms

Catchments	Models		FC	LP	ALFA	BETA	K	K4	PERC	CFLUX	MAXBAS
Alzette	ACCO	SO	284.83	0.26	0.06	0.65	0.02	0.01	0.16	0.04	10.96
		LF	356.34	0.46	0.10	0.42	0.02	0.00	0.14	0.10	13.48
		HF	414.48	0.19	0.30	0.49	0.00	0.03	0.97	0.01	8.51
	GA	SM	309.97	0.35	0.03	0.72	0.03	0.01	0.27	0.01	11.45
		LF	255.11	0.46	0.07	0.98	0.03	0.01	0.23	0.05	12.62
		HF	338.84	0.56	0.06	0.95	0.01	0.02	0.89	0.00	8.37
	NSGA-II	LF	253.24	0.16	0.07	0.54	0.02	0.00	0.13	0.00	9.49
		HF	253.25	0.34	0.07	0.52	0.02	0.01	0.14	0.00	9.54
Leaf	ACCO	SM	272.11	0.29	0.30	1.57	0.27	0.26	2.27	0.62	6.04
		LF	303.49	0.14	0.42	1.14	0.08	0.04	0.47	0.91	4.94
		HF	230.27	0.16	0.62	1.08	0.08	0.28	0.00	0.96	5.99
	GA	SM	349.80	0.64	0.65	2.29	0.07	0.14	0.65	0.97	5.99
		LF	313.84	0.22	0.26	1.24	0.14	0.05	0.66	1.00	5.06
		HF	285.21	1.00	0.72	1.91	0.05	0.26	1.83	0.99	6.00
	NSGA-II	LF	301.88	0.36	0.37	1.95	0.14	0.24	1.07	0.89	5.57
		HF	274.26	0.90	0.45	2.27	0.15	0.26	1.24	0.85	5.86
Bagmati	ACCO	SM	371.88	0.67	0.07	1.01	0.49	0.14	7.78	0.20	2.57
		LF	448.05	0.79	0.13	1.04	0.37	0.06	7.88	0.29	2.53
		HF	445.04	0.65	0.08	1.05	0.46	0.23	7.79	0.67	2.94
	GA	SM	430.26	0.59	3.85	1.04	0.50	0.08	8.00	0.01	2.99
		LF	301.36	0.83	0.22	1.05	0.22	0.06	7.99	0.10	2.39
		HF	453.50	0.58	0.07	1.23	0.49	0.03	0.05	0.00	2.91
	NSGA-II	LF	364.46	0.75	0.16	1.05	0.33	0.07	7.94	0.11	2.48
		HF	370.66	0.66	0.10	1.06	0.37	0.10	7.17	0.17	2.74

SM is single hydrological model (single optimal model-optimized by a single-objective optimization algorithm based on the classical *RMSE*); LF and HF are low flows model and high flows model (optimized by a single-objective optimization algorithms and multi-objective optimization based on the $RMSE_{LF}$ and $RMSE_{HF}$).

3.6   Results and discussion

3.6.1   Fuzzy committee models

Fuzzy committee models were tested in the three catchments, namely Alzette in Luxemburg, Leaf in the USA, and Bagmati in Nepal. The experimental design follows the one used in an earlier study (Fenicia et al., 2007), where the Alzette catchment was considered, and only calibration data were considered for building the models without further validation. This study presents two additional catchments (Leaf and Bagmati) with both calibration and verification periods, and compares the overall model performance when using different weighting schemes for objective functions (Figure 3-1) and different membership functions (Figure 3-2). The first two months of calibration data are considered as the warming-up period in the Leaf and Bagmati catchments. However, for the Alzette catchment, we used the hourly data set of only one year for the calibration period and one year for verification. We allocated 168 hours of data for the model warm-up period to compensate for the lack of data in the calibration set.

The ranges of HBV model parameters for optimization are given in Table 3-4. We produced the specialized models (the best single model specialized on high flows and low flows), which are optimized by multi- and single-objective optimization algorithms. The identified best sets of parameters for different models are given in Table 3-5. It is worth mentioning only one model structure (HBV) is used, but the use of several calibration methods and several error functions result in several model parameterizations, or instantiations. For simplicity, we will be mentioning several models, meaning actually the same model structure but with several parameterizations.

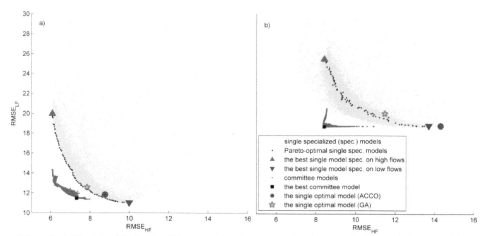

Figure 3-4. The identified sets of Pareto-optimal parameterizations of single specialized models (optimized by NSGAII), committee models, and single optimal models, calibrated by ACCO and GA, for the Leaf catchment. (a) calibration data set; (b) verification data set (model parameterizations from (a) are used).

We present the results of calibration using several optimization algorithms: NSGAII, GA and ACCO. The reasons for using different optimization algorithms for calibration are: 1) the GA initially used appeared to be quite slow (in terms of the required model runs). Consequenly, we decided to test the use of faster algorithms as well; 2) to cross-check one against the another, since they both use randomization of the initial population, and this affects the results. In each experiment, a committee model is compared with the single optimal model, which is calibrated by two different single-objective optimization algorithms (GA and ACCO). The best single models specialized for low and high flows, respectively (found by NSGA-II, GA, and ACCO), are used in the committee model. The points denoted "committee models" correspond to the model parameterizations generated during the exhaustive search for the best γ and δ, ensuring the lowest $RMSE$. We also tested a committee model that was built by combining the specialized models, and compared it against the single optimal model for all catchments. In Table 3-6 and Table 3-7, these models are denoted as $Q_c(ACCO)$, $Q_c(GA)$, and $Q_c(NSGAII)$. Interestingly, the committee model proved to be better on both objective functions ($RMSE$ and NSE) than the single optimal models, for all case studies.

The graphs of Pareto-optimal of single specialized models (calibrated by NSGAII), committee models, and single optimal models (calibrated by ACCO and GA) for the Leaf catchment are shown in Figure 3-4. In this figure, one may see that the 20205 single specialized models presented as they were generated during the multi-objective optimization process, by using the *WStype* I weighting scheme. The 70 model parameterizations identified as the best (Pareto-optimal) are presented as well (indicated by the darker points). The Pareto-optimal models in calibration are not necessarily the best in verification, which is, of course, no surprise. What is important to stress, however, is that their performance is not too much lower than that for the calibration data. Among the Pareto-optimal models, the best single model specialized on low flows achieved 11.01 in $RMSE_{LF}$ (all errors are in m^3/s), and it can be seen from Figure 3b that this model is also very close to being the best in verification with $RMSE_{LF}$ of 18.33. However, the best single model specialized on high flows ($RMSE_{HF}$ is 6.08) is not too bad but not the best in verification (Its $RMSE_{HF}$ is 8.42 and, in Figure 3b, it is easy to see that there are many models with lower $RMSE_{HF}$ values)

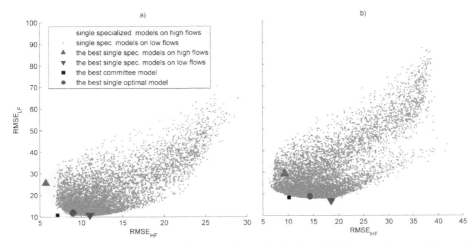

Figure 3-5. The identified sets specialized models (optimized by ACCO), committee models, and single optimal models, calibrated by ACCO in Leaf catchment. (a) calibration data set; (b) verification data set (model parameterizations from (a) are used).

The committee model resulted in the classical $RMSE$ of 15.63 in calibration and 25.23 in verification. To represent this model in Figure 3-4 (solid square), we had to calculate it for the corresponding $RMSE_{LF}$ and $RMSE_{HF}$ values, and we did this using the same type of weighting scheme (WStype I) that was used in the calibration of specialized models. In the same way, we presented the single optimal models identified using the two single objective optimization methods (ACCO and GA are represented by a circle and a star, respectively). It can be seen that the committee models are closer to the ideal point than the other single optimal models, and this means that the committee model's performance is highest among any of the single models.

The performances of the best single models specialized on high and low flows ($RMSE_{HF}$ and $RMSE_{HF}$) on various catchments are presented in Table 3-6. However, again, it should be noted that $RMSE_{LF}$ and $RMSE_{HF}$ cannot be compared, since they use different formulas. $RMSE_{LF}$ values are even higher than those of $RMSE_{HF}$. The reason is that the number of low flows is much higher than that of high flows, and the denominator (total number of observations) in both formulas is the same.

In the Leaf catchment, we tested all possible combinations of different weighting scheme types and classes of membership functions. These results are presented in Table 3-7. Noticeably, all committee models improved their performances in verification in comparison with the single hydrological models, which were optimized by single objective optimization. However, on the other two catchments, Alzette and Bagmati, the number of experiments was smaller, but, in all of them, the committee models demonstrated the highest performance on both calibration and test data sets.

Table 3-6 reports the performance of committee models and single-optimal models calibrated by ACCO and GA for each catchment. For the Bagmati catchment, the *RMSE*, calculated by a single optimal model in calibration, is 101.01 m³/s, and verification is 112.42 m³/s. However, it can be noticeably improved by the committee models and obtained values around 94.16 - 95.96 m³/s in calibration and 109.38 - 110.29 m³/s in verification. The *RMSE* of the single model produced 26.76 m³/s in the verification period for the Leaf catchment. However, when new types of weighting and membership functions were used, *RMSE* dropped to 23.41 m³/s (see Table 3-7).

The plots for the committee models that are built from the combination of the two specialized models for high and low flows, with respect to the hydrograph simulations, are illustrated in Figure 3-6. It can be observed that the committee model combines the best features of the specialized models.

Table 3-6. The performances of single optimal models (optimized based on classical *RMSE*) and committee models of various catchments. Committee models are assembled by a combination of the weighting scheme *WStype* I and membership function *MFtype* A.

Catchments	Models	Membership function		$RMSE_{HF}$		$RMSE_{LF}$		RMSE		NSE	
		δ	γ	Cal.	Ver.	Cal.	Ver.	Cal.	Ver.	Cal.	Ver.
Alzette	Qs (ACCO)	n a	n a	0.97	1.23	2.11	1.71	2.37	*2.39*	0.88	0.86
	Qs (GA)	n a	n a	0.99	1.2	2.03	2.01	*2.31*	2.42	*0.89*	*0.88*
	Qc (ACCO)	0.50	0.25	0.53	0.81	1.65	1.48	2.10	2.06	0.91	0.89
	Qc (GA)	0.60	0.40	0.56	0.78	1.71	1.35	2.19	2.15	0.90	0.90
	Qc (NSGAII)	0.50	0.30	0.50	0.86	1.48	1.48	**1.99**	2.07	0.93	0.89
Leaf	Qs (ACCO)	n a	n a	8.97	14.31	11.84	18.81	17.56	26.76	0.87	0.83
	Qs (GA)	n a	n a	7.96	11.70	11.71	19.64	*17.36*	*26.58*	*0.88*	*0.84*
	Qc (ACCO)	0.39	0.37	5.88	9.00	10.55	18.74	**15.63**	25.23	**0.91**	0.85
	Qc (GA)	0.51	0.50	5.82	8.91	10.86	19.38	15.76	24.88	0.90	0.85
	Qc (NSGAII)	0.50	0.49	5.61	9.05	10.85	18.59	16.05	23.86	0.90	**0.88**
Bagmati	Qs (ACCO)	n a	n a	29.55	85.33	38.84	87.54	*101.01*	*112.42*	*0.87*	*0.90*
	Qs (GA)	n a	n a	32.42	86.18	34.16	92.38	101.69	116.72	0.86	0.88
	Qc (ACCO)	0.61	0.49	40.09	69.31	72.69	65.09	95.96	**109.38**	0.87	0.90
	Qc (GA)	0.57	0.47	18.93	39.75	77.29	74.55	94.39	110.29	**0.89**	**0.91**
	Qc (NSGAII)	0.50	0.47	26.94	48.34	82.42	81.67	**94.16**	109.72	0.87	**0.91**

Note: In Table 3-6, Qs (ACCO) and-Qs(GA) indicate single hydrological models (SMs), Qc(ACCO), Qc(GA), and Qc(NSGAII) are committee models (CMs), (Bold indicates best CMs, italics is best SMs and highlighted grey is best model for a catchment, n.a. is not available)

These experiments have led to one important observation related to using the weighting function for objective functions (Figure 3-1 and Equation 3-7) in calibration of specialized models; that is, that the-quadratic function used earlier (Fenicia et al., 2007) was, in fact, the first indication that it will allow for distinguishing the low and high flows. In our latest experiments, it appeared, quite expectedly, that other functions (for example, cubic) may work better during the calibration period.

Table 3-7. Performance of committee models for the Leaf catchment, with possible combinations of the various weighting schemes and membership functions.

Models	Weighted function		Membership function			$RMSE_{HF}$		$RMSE_{LF}$		RMSE		NSE	
	WStype	N	MFtype	δ	γ	Cal.	Ver.	Cal.	Ver.	Cal.	Ver.	Cal.	Ver.
Qs (ACCO)	n a	n a	n a	n a	8.97	14.31	11.84	18.81	n a	17.56	26.76	0.87	0.83
Qs (GA)	n a	n a	n a	n a	7.96	11.70	11.71	19.64	n a	*17.36*	*26.58*	*0.88*	*0.84*
Qc (ACCO)	I	2	A	0.39	0.37	5.88	9.00	10.55	18.74	15.63	25.23	0.91	0.85
	II	2	A	0.45	0.44	7.27	10.47	9.71	17.86	16.01	24.38	0.89	0.85
	III	2	A	0.65	0.14	1.02	3.23	9.56	17.26	**15.60**	24.52	0.92	0.88
	IV	2	A	0.56	0.55	1.13	3.00	11.47	19.64	16.20	25.68	0.87	0.86
	I	2	B	0.39	0.38	5.88	9.00	10.55	18.74	15.63	25.26	0.90	0.85
	II	2	B	0.45	0.44	7.27	10.47	9.71	17.86	16.03	24.38	0.89	0.85
	III	2	B	0.94	0.15	1.02	3.23	9.56	17.26	15.67	24.72	0.92	0.88
	IV	2	B	0.56	0.55	1.13	3.00	11.47	19.64	16.20	25.68	0.87	0.85
Qc (GA)	I	2	A	0.51	0.50	5.82	8.91	10.86	19.38	15.76	24.88	0.90	0.85
	II	2	A	0.66	0.14	7.36	9.84	9.81	18.93	16.13	25.81	0.92	0.86
	III	2	A	0.99	0.16	0.99	2.77	9.67	18.36	16.53	24.67	0.92	0.87
	IV	2	A	0.99	0.30	1.01	2.85	11.50	19.84	16.60	23.96	0.92	0.87
	I	2	B	0.99	0.15	5.82	8.91	10.86	19.38	16.30	25.56	**0.93**	0.86
	II	2	B	0.87	0.16	7.36	9.84	9.81	18.93	16.22	25.58	0.93	0.87
	III	2	B	0.99	0.31	0.99	2.77	9.67	18.36	16.47	24.34	0.92	0.87
	IV	2	B	0.99	0.42	1.01	2.85	11.50	19.84	16.55	24.06	0.92	0.87
	I	1	B	0.42	0.41	9.43	13.06	12.44	20.56	15.96	24.04	0.91	0.89
	I	3	B	0.99	0.23	3.64	6.95	9.30	17.08	16.50	25.53	0.91	0.86
Qc (NSGAII)	I	2	A	0.50	0.49	5.61	9.05	10.85	18.59	16.05	23.86	0.90	0.88
	II	2	A	0.50	0.49	7.31	9.98	9.70	17.45	15.71	23.85	0.91	0.88
	III	2	A	0.86	0.47	1.08	2.86	10.13	17.76	17.36	**23.41**	0.91	0.90
	IV	2	A	0.86	0.45	1.08	3.10	11.62	20.01	16.76	23.97	0.90	0.87
	I	2	B	0.50	0.29	5.61	9.05	10.85	18.59	16.45	23.96	0.90	0.88
	II	2	B	0.50	0.15	7.31	9.98	9.70	17.45	16.71	23.95	0.91	0.88
	III	2	B	0.99	0.49	1.08	2.86	10.13	17.76	17.29	23.46	0.91	0.91
	IV	2	B	0.99	0.46	1.08	3.10	11.62	20.01	16.71	23.97	0.91	0.88
	I	1	A	0.38	0.36	9.59	12.76	12.91	20.46	16.58	23.86	0.91	**0.91**
	I	3	A	0.50	0.49	4.19	7.88	9.60	17.07	15.96	23.79	0.90	0.88

Note: The value of α =0.75 used in *WStype* II, III and IV (Bold = best CMs, italics = best SMs)

The performances of the committee models that were built from the combination of the two local models for high flow and low flow, with respect to the hydrograph simulations, are presented in Figure 3-6. It can be visually observed that the committee model features the best characteristics of both specialized models.

One potential issue related to the scaling of weighting function for objective functions used in fuzzy committee model is worth mentioning that Equation (3-3) used in all *WStype* that involve $Q_{o,max}$. This may become a certain problem in operation (verification). $Q_{o,max}$ is the maximum for calibration data, but this, of course, does not guarantee that it will not be superseded in the future when the model is in actual operation (or when simulating the operation by using verification data). The quadratic function will still handle values above 1, but if the calibration maximum is exceeded considerably, then the high flow will be given disproportionally higher weights, and low flows disproportionally lower weights. A solution may lie in using a slightly wider range for scaling.

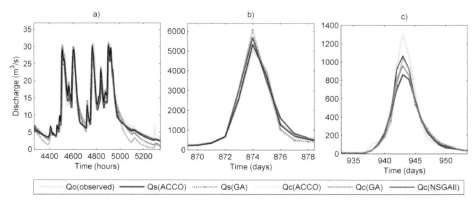

Figure 3-6. A fragment of hydrograph generated from various models, Qo is observed discharge, Qs indicates model identified by single-objective optimization (ACCO and GA), Qc is committee model (ACCO, GA and NSGAII), (a) Alzette (31/01/2002 08:00:00 - 18/03/2002 03:00:00); (b) Bagmati (20/5/1990-28/5/1990); and (c) Leaf (13/02/1960 - 08/03/1960

3.6.2 States-, inputs-, and outputs-based committee models for Brue, Bagmati, and Leaf

The predictive capability of inputs, states and outputs-based committee models are tested on the three catchments (Bagmati, Brue, and Leaf) with both calibration and verification periods. The first two specialized models (the best single model specialized on high flows and low flows) are optimized by single-objective optimization algorithms with objective functions (Figure 3-1 a), and then combined using weights Equations 3-14 to Equations 3-17. The overall model performances are compared with a single optimal model and fuzzy committee model.

The observed and simulated hydrographs of the best combined cases of specialized models for all three catchments are plotted in Figure 3-7 to Figure 3-9 (the hydrograph generated from various committee models). Each of the three catchments exhibited different hydrological behaviour, which was reflected in the shape of the hydrograph. The six graphs demonstrated the ability of nine different committee models to capture most of the hydrograph features in calibration and verification. In plots Figure 3-7 to Figure 3-9, MSO is the single hydrological model (single optimal model-optimized by a single-objective optimization algorithm based on the classical $RMSE$); MLF and MHF are the low-flows model and the high-flows model, respectively (optimized by a single-objective optimization algorithms based on the $RMSE_{LF}$ and $RMSE_{HF}$); MFM is a fuzzy committee model; MSV1, MSV2, MSV3 are states-based committee models (weights based on soil moisture level, upper tank level, and lower tank level, respectively); MBF1, MBF2, MBF3 are outputs-based committee models (weights based on quick flow, base flow, and simulated streamflows, respectively); MBI1 and MBI2 are inputs-based committee models (weights based on precipitation and evapotranspiration, respectively).

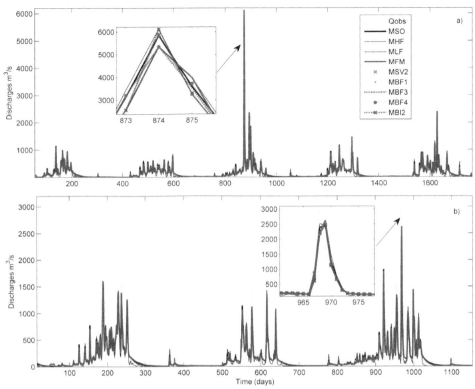

Figure 3-7. Hydrographs generated from the various committee models for the Bagmati catchment: (a) calibration and (b) verification

The committee model MFB3 demonstrated an excellent performance in reproducing the observed hydrograph at peak for the Brue and Leaf catchments (Figure 3-8 and Figure 3-9), and to some extent in the Bagmati catchment (Figure 3-7), but features, such as rising limb, are better generated by this model than the individual peak values.

Table 3-8. The identified set of model parameters by ACCO for Brue catchment

Parameters	SO	HF	LF
FC	163.56	172.32	191.88
LP	0.61	0.59	0.62
ALFA	1.70	1.63	1.38
BETA	1.89	1.81	1.99
K	7.4E-04	1.0E-03	1.6E-03
K4	3.2E-03	3.0E-03	4.1E-03
PERC	8.8E-02	5.5E-02	0.09
CFLUX	0.04	0.02	0.04
MAXBAS	12.09	12.81	10.93

SO: Single optimal model; LF: Low-flow model; HF: High-flow model

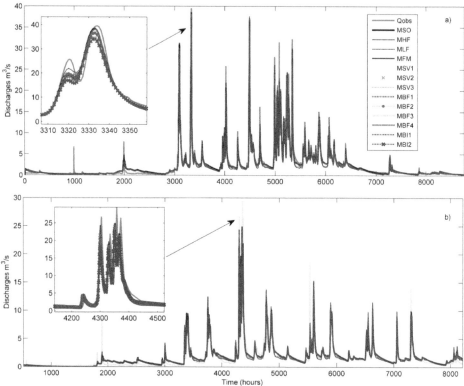

Figure 3-8. A hydrograph generated from various committee models for the Brue catchment. (a) calibration and (b) verification

Table 3-9 shows the overall performance in *RMSE* and *NSE* for all committee models, and for the single-optimal model in calibration and verification period for three catchments. The results in calibration and validation for the different committed models significantly differ from each other. *RMSE* values of fuzzy committee models (MFM) for calibration and verification are higher among all models. In Bagmati, committee models MFM, MSV2, MBF1 and MBF3 performed better than MSO, both in calibration and in verification. MBI1 improved the performance in calibration but not in validation. MSV1 model performance is the worst among all models. The value of *RMSE* obtained by the MBF3 represented a significant improvement in the calibration period for Brue but insignificant progress in verification. There were no significant differences among the NSE in verification. Models MFM, MSV1, MSV2, MSV3, and MBF3 performed better than MSO for the Leaf catchment. MFM and MBF3 are best among all models in calibration and in verification, respectively. *RMSE* of the MSV1 is almost equal to MSO in calibration, which is greater than the MFM, MVS2, and MBF1 but is still best in verification.

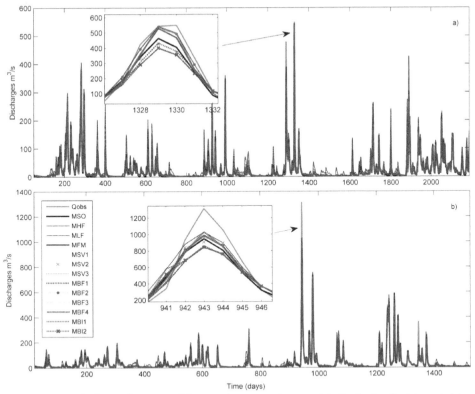

Figure 3-9. A hydrograph generated from various committee models for the Leaf catchment. (a) calibration and (b) verification

The overall results show that most committee models performed better than the single model, and, among all of them, MFM and MBF3 are the best. Compared with other committee models, model MOD did not perform well in its performances. In this model, the weights are calculated from the SMA (soil moisture accounting) of the model that balanced the low- and high-flow models (these two models are calibrated by log transformed for low flows and *RMSE* for high-flow), while these weights help to increase or decrease flow that affects the shape of the hydrograph. In MFM, fuzzy membership function switches to a smooth transition between the boundaries of specialized models, which does not allow additional flow into the system.

It is useful to mention that overfitting the problem (typically addressed in data-driven modelling) may lead to a decrease of accuracy in operation (which can be detected during validation), and there are several ways to address this issue. For example, trying to limit the complexity of a model, and/or using cross-validation data set to control the calibration process. The committee models include weights (MFM includes a number of parameters) that increase accuracy of the overall model. However, this contributes to increasing the complexity of the model and may lead to overfitting.

Table 3-9. The performances of single optimal models and committee models (*RMSE* and *NSE*) of various catchments

		Bagmati				Brue				Leaf			
		RMSE		NSE		RMSE		NSE		RMSE		NSE	
SN	Models	Cal.	Ver.	Cal.	Ver.	Cal.	Ver.	Cal.	Ver.	Cal.	Ver.	Cal.	Ver.
1	MSO	101.01	112.42	0.87	0.89	0.79	0.99	**0.96**	0.83	17.56	26.76	0.88	0.90
2	MFM	**95.66**	**109.38**	**0.89**	**0.90**	0.78	0.97	**0.96**	0.83	**15.63**	25.23	**0.89**	0.91
2	MOD*	105.45	113.20	0.85	0.82	0.87	0.99	0.95	0.83	18.51	27.98	0.87	0.89
4	MSV1	107.61	124.25	0.84	**0.77**	0.88	0.97	0.94	0.83	17.55	24.98	0.87	0.91
5	MSV2	99.17	110..25	0.88	0.83	0.80	0.97	0.95	0.83	15.81	25.00	**0.89**	0.91
6	MSV3	104.21	115.79	0.87	0.80	0.85	**0.96**	0.95	**0.84**	20.03	29.59	0.83	0.88
7	MBF1	99.57	111.15	0.88	0.83	0.83	0.98	0.95	0.83	17.33	28.32	0.87	0.89
8	MBF2	104.21	115.79	0.87	0.80	0.85	**0.96**	0.95	**0.84**	20.03	29.59	0.83	0.88
9	MBF3	97.88	111.75	0.88	0.82	**0.74**	0.97	**0.96**	0.83	15.90	**24.30**	**0.89**	**0.92**
10	MBI1	105.47	118.44	0.87	0.79	0.83	0.98	0.95	0.83	17.50	28.07	0.87	0.89
11	MBI2	100.71	114.27	0.87	0.83	0.83	**0.96**	0.95	0.83	17.37	27.94	0.87	0.89

*(Oudin et al., 2006)

3.7 Summary

This chapter provides a comparative assessment of the various committee models that are assembled by different model combination methods to improve the prediction of catchment hydrological responses. A committee model consists of the specialised models (separately calibrated for each regime with different objective function), which are combined using a proper weighting method. The resulting model is subsequently tested on verification data and compared with other models based on objective function *RMSE* and *NSE*. The differently-parameterized models show strengths in encapsulating different characteristics of the catchment responses. Different weighting schemes for objective functions and membership functions are compared in terms of the overall model performance.

Chapter 4
Hybrid committees of hydrological models

This chapter presents hybrid models built from a combination of the conceptual and data-driven hydrological models. These two individual models are responsible for a particular aspect of flows, this chapter describes their combination into one new model to form a hybrid committee model. We tested the performance of hybrid committee models by using several weighting schemes in objective functions for calibration of individual specialized models, and final models are compared with single hydrological models (HBV and ANN) for the Bagmati and Leaf catchments. The results show that hybrid committee model can significantly improve the performances of the model.[2]

4.1   Introduction

Conceptual and data-driven models each have a number of advantages and drawbacks. Therefore, a new model can be built, by combining the best features of these models into what is known as a "hybrid model". The idea of hybrid modelling is not new in the hydrological community, as discussed by some authors in the literature (Abrahart and See, 2002; Corzo and Solomatine, 2006; Corzo et al., 2009). Various types of hybrid models have been proposed; some examples are: wavelet-ANN model (Anctil and Tape, 2004); chaotic-ANN model (Karunasinghe and Liong, 2006); semi-distributed processed-based; and the ANN model (Corzo et al., 2009).

Hybrid models can be classified in three ways (Corzo, 2009): (i) data-driven models with incorporated hydrological knowledge (e.g., the separation of input space or the identification of processes and regimes of outputs based on hydrological knowledge); (ii) process-based models using data-driven techniques, or with some components replaced by data-driven models to solve complex processes in a physically-based model; and (iii) data-driven models used in parallel (e.g., ensembles) or sequentially (e.g., in a data assimilation setup).

This study implements the first type of hybrid model, with this model initially building conceptual and data-driven models. These models specialized on high flows and low flows, respectively, which are combined by fuzzy membership function. The conceptual models for the low flows of the catchment (hydrograph) are typically difficult to calibrate, and one possibility here is to use a data-driven model (ANN) for low-flows. The high-flows can be modelled by a conceptual hydrological model. (Note that the data-driven (low-flow) component does not have hydrological states, whereas the conceptual model does.)

---

[2] Kayastha, N. and Solomatine, D. P., (2013). Combinations of specialized conceptual and neural network rainfall-runoff models: comparison of performance, *Proc. Geophysical Research Abstract, European Geosciences Union.* Vol. 15, EGU 2013-9022.

This study tests the performance of a fuzzy hybrid committee models for two catchments (Bagmati and Leaf), and used the same weighting scheme in objective functions for calibration of specialized models of the HBV hydrological model and ANN model. [This is presented in the previous chapter (see, Figure 3-1a)].

4.2   Low flows simulation

Hydrological models often poorly simulate low flows, since these models are traditionally designed to simulate the runoff response to rainfall, and it is difficult for them to learn the rainfall-runoff relationships. Furthermore, the values of low flows of the hydrograph are considerably lower than those of the high flows during storm events, and have the most values that are close to zero. The criteria used for the evaluation of simulation depend on the type of hydrological regime or the type of model application. Most of the performance criteria are based on the mean model squared error, a criterion which is known to place greater emphasis on high flows simulation. Pushpalatha et al, (2012) presented different criteria for evaluating the simulation of low flows in hydrological models. They found that the criterion calculated on inverse flows is better suited for the evaluation in low-flow conditions. The predictability of a hydrological model for low flows is relies on a description of the relationship between the surface water and groundwater processes in low-flow conditions. In reality, low-flow prediction/forecasting with hydrological models is very complex because processes conditioning low flows involve various inherently complex predictors (e.g., climatic factors). In addition, it also depends on the region and the season. One solution could be to use a data-driven model, e.g., an ANN, model because in hydrological modelling, data-driven moels have been shown to be excellent flow predictors due to their high nonlinearity, flexibility, and without much prior knowledge about catchment behaviours and flow processes. This study proposes to use the ANN model instead of the HBV model for low flows in committee models.

4.3   ANN models specialized on low flows

The motivation behind using ANN is that this is a powerful tool for the modelling of non-linear systems, such as river streamflows. It allows building a simple model, that is decomposed from the larger problem, and the modelling of decomposed streamflows performs better than using one single model (Jain and Srinivasulu, 2006; Corzo and Solomatine 2007).

A number of specialized rainfall runoff models were built instead of using only one single model. The discussion on *specialized model* can be found in Chapter 3, Section 3. The input variables for artificial neural network (ANN) models that specialized on low flows are selected based on correlation and mutual information analysis between the input and output variables, as described in Chapter 2. The variables selected for the ANN models are shown in Equation 4-1. Figure 4-1 presents the autocorrelation and cross-correlation results for the Bagmati and Leaf catchments. The peaks of both the cross correlation and mutual information are observed at 1 hour and 3 hours, respectively. Along with this analysis, several models with different combinations of variables have been selected as well, but here we have only shown the model that results in the lowest error. Hydro-meteorological time series of more recent data series at different previous time points are used as model inputs to predict the original data series at the current time moment. For daily time series, the past data, from one day prior to a few days prior, are typically used as model inputs.

The ANN rainfall runoff models specializing on low flows for the Bagmati and Leaf catchment are given below:

$$Q_{LF} = f(R_{t-1}, R_{t-1}, Q_{t-1}) \qquad (4\text{-}1)$$

$$Q_{LF} = f(R_{t-3}, R_{t-4}, R_{t-5}, Q_{t-1}, \Delta Q_{t-1}) \qquad (4\text{-}2)$$

where R_{t-1} is rainfall of the previous day, Q_{t-1} is previous flow, and ΔQ_{t-1} is the change in flow (flow derivative). The flow data series at one day prior is always selected as one of the inputs because of the usual high lag-1 autocorrelation (Kisi, 2008). This selection principle used of single-step-ahead prediction in which each single step ahead can predict the next one-day datum. However, this study predicts current stage from one-step previous input (simulation mode).

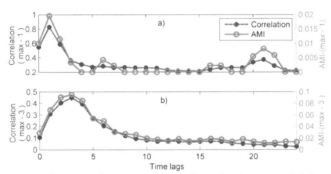

Figure 4-1. Plots of correlations and average mutual information between rainfall and low-flows (a) Bagmati catchment and (b) Leaf catchment

4.4 Committee of ANN and HBV for Bagmati and Leaf

Specialized HBV and ANN rainfall runoff models are built to reproduce catchment hydrological responses, which are specialized on distinctive regimes (high flows and low flows) and are combined using an appropriate combining scheme. The combining scheme was used to weight the contributions of each specialized model, which makes use of a fuzzy attribution of weights, and where the outputs of the models are multiplied by the weights that depend on the value of flow and then normalized. This is shown in Equation 3-9, and the descriptive figure is given in Figure 3-2. Several committee models can be formed by combining the two specialized models using a fuzzy committee. This study tests three different committee models, which are built by a different combination of specialized models, namely (i) low-flow HBV model and high flow HBV model; (ii) low-flow ANN model and high flow ANN model; and (iii) low-flow ANN model and high-flow HBV model. The committee model Q_c is calculated by combination sets of δ and γ, which are selected within given intervals, and the performance measure is calculated by root *RMSE* and *NSE*. The membership functions are subject to the optimization of parameters (γ, δ) and shapes, for the best combination.

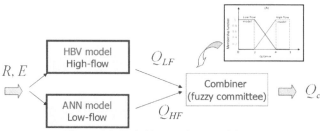

Figure 4-2. Hybrid committee model

4.5 Results and discussion

The performances of all hybrid committee models are presented in Table 4-1. The results of the hybrid models show that their performance is best among all committee models. The use of the ANN technique is effective in prediction, but not necessarily in terms of the modelling efficiency. The reason is that the ANN technique built model is based on relationship inputs and outputs rather than on hydrological processes. The hybrid committee approach uses two techniques of modelling (separately used within the two models), which yield a better model performance.

Table 4-1. Performances of the hybrid model and single hydrological models

Models	Specialised models	Weighted function N	Membership function δ	γ	RMSE Calibra-tion	RMSE Verifi-cation	NSE Calibra-tion	NSE Verifi-cation
(a) Bagmati								
Q_{HBVs}	n a	n a	n a	n a	101.01	112.42	0.87	0.82
Q_{ANNs}	n a	n a	n a	n a	68.27	110.89	0.94	0.82
$Q_{CHBV+ANN}$	$HBV_{HF} + ANN_{LF}$	2	0.73	0.01	72.69	105.88	0.93	0.84
$Q_{CHBV+ANN}$	$HBV_{HF} + ANN_{LF}$	3	0.94	0.02	72.40	103.98	0.93	0.84
Q_{CHBV}	$HBV_{HF} + HBV_{LF}$	2	0.64	0.47	95.66	111.13	0.89	0.82
Q_{CHBV}	$HBV_{HF} + HBV_{LF}$	3	0.51	0.46	94.38	109.59	0.89	0.82
Q_{CANN}	$ANN_{HF} + ANN_{LF}$	2	0.37	0.11	66.75	109.59	0.94	0.82
Q_{CANN}	$ANN_{HF} + ANN_{LF}$	3	0.88	0.75	68.30	109.07	0.94	0.83
(b) Leaf								
Q_{HBVs}	n a	n a	n a	n a	17.56	26.76	0.88	0.90
Q_{ANNs}	n a	n a	n a	n a	11.49	24.90	0.94	0.91
$Q_{CHBV+ANN}$	$HBV_{HF} + ANN_{LF}$	2	0.73	0.49	12.95	18.92	0.93	0.95
$Q_{CHBV+ANN}$	$HBV_{HF} + ANN_{LF}$	3	0.81	0.02	12.12	19.65	0.94	0.95
Q_{CHBV}	$HBV_{HF} + HBV_{LF}$	2	0.39	0.37	15.63	25.23	0.89	0.91
Q_{CHBV}	$HBV_{HF} + HBV_{LF}$	3	0.51	0.49	15.96	23.79	0.90	0.92
Q_{CANN}	$ANN_{HF} + ANN_{LF}$	2	0.66	0.55	11.26	25.11	0.94	0.91
Q_{CANN}	$ANN_{HF} + ANN_{LF}$	3	0.64	0.5	10.51	20.96	0.95	0.94

The value of *RMSE* calculated for single optimal model in calibration for the Bagmati catchment, is 101.01 m^3/s and verification is 112.42 m^3/s However, these values can be noticeably improved by the committee models (ANN), attaining around 66.75- 68.30 m^3/s in calibration and 109.07 - 109.59 m^3/s in verification. The best committee models (hybrid of HBV and ANN) obtained values around 72.40 - 72.69 m^3/s in calibration and 103.98 - 105.88 m^3/s in verification. The *RMSE* of the single model produced 26.76 m^3/s in verification period for the Leaf catchment. However, when a hybrid of ANN and HBV model was used, *RMSE* dropped to 18.92 m^3/s.

Comparisons of the performances of different models in calibration and verification are presented in Table 4-1 for Bagmati and Leaf, in this table, the best values of *RMSE* and *NSE* were shown by hybrid models $Q_{CHBV+ANN}$, that is combination of specialized models (low-flow model by ANN and high-flow model by HBV). The visual plots of the hybrid committee models, with respect to the hydrograph simulations, are represented in Figure 4-3 to Figure 4-6. It can be observed that the hybrid committee model combines the best features of the specialized models. The overall results show that the hybrid models (combination of HBV and ANN) have the highest accuracy among all models.

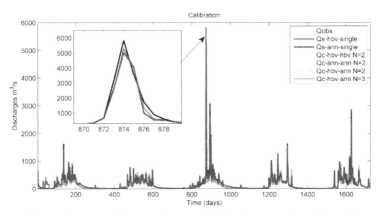

Figure 4-3. Hydrograph generated from various models for the Bagmati catchment (calibration)

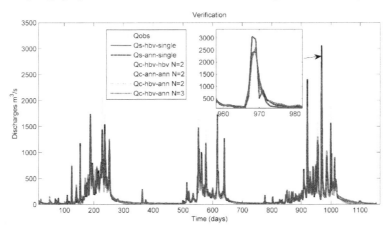

Figure 4-4. Hydrograph generated from various models for the Bagmati catchment (verification)

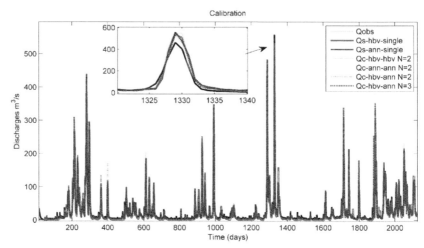

Figure 4-5. Hydrograph generated from various models for the Leaf catchment (calibration)

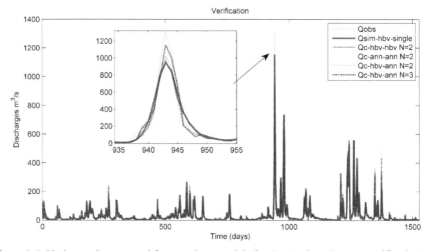

Figure 4-6. Hydrograph generated from various models for the Leaf catchment (verification)

A hybrid committee model can be used when a single hydrological model fails to identify catchment responses. The ANN model performs better than the traditional HBV hydrological model for daily runoff simulations over both catchments. The results show that the combination of HBV and ANN models can improve the performances (i. e., decrease the value of *RMSE*) by about 22% and 20% in calibration and verification respectively, in contrast to the committee model, that is, hybrid models achieve higher accuracy.

4.6   Summary

The ANN technique is a powerful tool for simulating the behaviour non-linear systems. In this study, the objective of the built hybrid model is to enhance model performance in streamflows modelling by integrating the artificial neural network (ANN) into it. ANN techniques were combined with a conceptual hydrological model in a committee. The proposed hybrid approach can be useful in developing efficient models of the complex, dynamic, and nonlinear rainfall-runoff process and can accurately predict the streamflows in the catchment. The best committee model was identified by the calibration data set and tested on the verification data set. The hybrid models outperformed the committee and single models on all case studies (Table 4-1). The results show that the daily streamflows predicted by the hybrid model have much better agreement with the observed data, while those simulated by the single hydrological model underestimate main peak-flows.

Chapter 5
Model parametric uncertainty and effects of sampling strategies

This chapter presents various sampling-based uncertainty analysis methods to quantify the uncertainty source from the parameters of the hydrological model. The uncertainty analysis methods differ with respect to sampling strategy (algorithm) that was used to generate the system responses. This study compares seven different sampling algorithms namely MCS, GLUE, MCMC, SCEMUA, DREAM, PSO and ACCO. The HBV conceptual hydrological model is set for Nzoia catchment as a case study. The results were compared and analyzed based on the shape of the posterior distribution of the parameters, uncertainty results on model outputs and sampling algorithm's ease of use.[3]

5.1 Introduction

The quantification of predictive model uncertainty is essential for decision-making processes (e.g., Beven and Binley, 1992; Vrugt et al., 2003; Liu and Gupta, 2007; Yang et al., 2008; Keating et al., 2010). Researchers seek to understand the process knowledge subject to considerable uncertainties from different sources of hydrological models. Over the past decade, the topic of exploring uncertainty from the parameters of hydrological models has received much attention in the hydrological community. Usually the model parameters are estimated by calibration, wherein the values of unknown model parameters are determined within the selected range by tuning until they match the observed predictions as closely as possible. The distributions and ranges of parameters are assumed to show the variation of parameters and, thus able to generate many model outputs. It is possible for various parameter sets within the chosen model structure yield the same effects in the model output because of the complex and highly nonlinear nature of the hydrological processes described in the model, further, changes in some parameters might be compensated by others (Bardossy and Singh, 2008). This effect is called an "equifinality" problem (Beven and Freer, 2001) and induces high uncertainties in the model predictions.

The estimation of hydrological model parameters is not always easy, and this problem has been explored in numerous studies (e.g., Bates and Campbell, 2001; Duan et al., 1992; Vrugt et al., 2003; Feyen et al., 2008; Beven, 2009). The main difficulties are related to searching problems of global parameters due to: (i) occurrence of multiple local optima, (ii) absences of

---

[3] Kayastha, N., Solomatine, D. P., Shrestha, D. L., van Griensven, A., (2013). Use of different sampling schemes in machine learning-based prediction of hydrological models' uncertainty, *Proc. Geophysical Research Abstract, European Geosciences Union*. Vol. 15, EGU 2013-9466.

Kayastha, N., Shrestha, D.L., Solomatine, D. P. (2011). Influence of sampling strategies on estimation of hydrological models uncertainty, *Proc. Geophysical Research Abstracts, European Geosciences Union*. Vol. 13, EGU 2011, 3781.

Kayastha, N., Shrestha, D. L., Solomatine, D. P. (2010). Experiments with several methods of parameter uncertainty estimation in hydrological modelling, *Proc. 9th International conference on Hydroinformatics*, China

prior information of joint probability distribution of parameters, which are often assumed to be independent, (iii) nonlinear interaction between parameters, and (iv) response surface of the selected objective function during optimization.
.

The estimation of parameters usually entails the searching for optimal parameters (calibration) of hydrological models. Modellers often fail to consider a subsequent model uncertainty prediction. However, both of these procedures (calibration and uncertainty analysis) relate are related (Blasone et al., 2008a; Laloy et al., 2010; Zhang et al., 2009). Therefore, the model predictions should be represented with respect to some confidence range (Gupta et al., 1998; Beven, 2006; van Grienven et al., 2008) so that the proper uncertainty can be quantified; the corresponding prediction limits on model outputs characterize the uncertainty of hydrological responses (Vrugt and Bouten, 2002; Feyen et al., 2007, 2008).

The approaches of uncertainty analysis differ in philosophy, underlying assumptions, understanding and quantification of different sources of uncertainty. Classification of the approaches to uncertainty analysis of hydrological models can be found e.g. in Montanari, (2011).

This chapter investigates the effects of sampling strategies on an estimation of parametric uncertainty in model predictions. The sampling strategy involves the choice of algorithm in sampling of parameters in the given range and distribution. Seven different methods were used to estimate the uncertainty of the conceptual hydrological HBV model for the Nzoia catchment in Kenya. These methods included Monte Carlo (MC) simulation (based on Latin hypercube simulation), generalized likelihood uncertainty estimation (GLUE by Beven and Binley, 1992), Markov Chain Monte Carlo (MCMC), shuffled complex evolution metropolis algorithm (SCEMUA by Vrugt et al., 2003), differential evolution adaptive Metropolis (DREAM by Vrugt et al., 2008), particle swarm optimization (PSO by Kennedy and Eberhart, 1995) and adaptive cluster covering (ACCO by Solomatine, 1999).

5.2 Comparison of parameter estimation and uncertainty analysis methods

Various approaches have been taken to analyse the uncertainty of hydrological models. Comparative studies of uncertainty analysis approaches are essential because they help to facilitate and broaden the knowledge concerning uncertainty study, improve the understanding of uncertain systems, and permit the determination of the prediction capability of different methods. The different approaches of parameter uncertainty estimation and their comparative studies can be found in the literature (see, Table 5-1). Most comparative studies are limited in parameter estimation (searching for optimum parameter set) and do not consider uncertainty estimation. A few studies have considered subsequent uncertainties in model outputs after parameter estimation, when compared during calibration (e.g., Vrugt et al., 2005; Vrugt et al., 2008; Zhang et al., 2009; Jin et al., 2010; Zhang and Zhao, 2012). However, these studies evaluated multiple sources of uncertainties (input, parameter and model structure). In contrast, very few studies have been made a concerning the comparison of different uncertainty estimation approaches for sources of uncertainty from the parameter of hydrological models.

Table 5-1. The uncertainty analysis methods compared by different authors

Sn	Year	Authors / Methods compared	Criteria for comparison	Optimization	Uncertainty	Model	Study catchment(s)
1	1997	Kuzcera (1997) — SCE, GA, MS, and Mq-N	Objective function second-order approximation to the response surface around the local minimum	Optimum parameter	X	mSFB Model (Boughton, 1984)	Chichester River catchment, in New South Wales, Australia
2	1999	Thyer et al. (1999) — SCE and SA-SX	Residual sum of square, empirical quantile plot of objective function, and number of function evaluations	Optimum parameter	X	mSFB model (Sumner et al., 1997)	Allyn River basin, and Scott Creek catchment, Australia
3	2002	Madsen et al. (2002) — SCE ,CAS and KBES	Water balance error (%), NS, RMSE, and Peak and low flow statistics including bias and RMSE.	Optimum parameter	X	NAM	Tryggevaelde catchment, Denmark
4	2004	Marshall et al. (2004) — AM, MHBC, MHBU and MHSS	Poster distribution of parameter, Mean, median, skewness of distribution , computation time, autocorrelation function for parameter		X	AWBM	Bass river catchment Australia
5	2005	Chen et al. (2005) — MP and SCE	Root mean square error (RMSE), root-mean-square of relative error (RR)	Optimum parameter	X	Tank model (Sugawara, 1995)	Shin-Fa Bridge Station catchment, Taiwan
6	2005	Vrught et al. (2005) — SODA and SCEMUA	Sum of weighted squared differences optimization	Prediction Interval		HYMOD	Leaf River
7	2006	Skahill and Doherty (2006) — PD_MS2 and SCE	NS			HSPF model (Bicknell et al., 2001) hydrologic model	Wildcat Creek watershed, USA
8	2007	Blasone et al. (2007) — SCE and PEST	Parameter convergence and correlation, MSE	Optimal parameter	X	MIKE SHE (Steady-state GW model, transient GW model and fully integrated model)	Karup catchment, Denmark

9	2007	Gattke and Schumann (2007) ———— GLUE and MG	NS	X	0.05- and 0.95- quantiles of the simulated discharge	HBV-96 model (Lump and Semi-distributed)	Meiningen catchment, Germany
10	2007	Lee et al. (2007) ———— SCEMUA and SCE	Simplest least square (SLS), Heteroscedastic Maximum likelihood estimator(HMLE) and Modified index of agreement (MIA)	Optimal parameter	Uncertainty boundary estimated by SCEMUA	KsEdgFC2D (Tachikawa et al., 2004)	Kamishiiba catchment, Japan
11	2008	Blasone et al. (2008a) GLUE based SCEMUA and GLUE based LHS	Likelihood function	Optimal parameter	Percentage of observations, uncertainty intervals	HYMOD, NAM and SAC-SMA	Tryggevaelde catchment Denmark and Leaf River catchment, USA
12	2008	Smith and Marshall (2008) ———— AM, DRAM, DEMC	Likelihood function	Optimal parameter		PDM	TCEF, USA
13	2008	van Greinsven et al. (2008) ParaSol and SUNGLASSES	Global Optimisation Criterion(GOC) (van Griensven & Meixner, 2007) and BIAS	X	Confidence interval	SWAT	Honey Creek, Ohio, USA
14	2008	Vrugt et al. (2008) ———— DREAM and GLUE	RMSE, CORR, BIAS, POC, Width of uncertainty interval	X	Uncertainty intervals	HYMOD	Leaf river, USA and French broad watershed
15	2008	Yang et al. (2008) ———— GLUE, ParaSol, SUFI-2, MCMC, and IS.	Best parameter, correlations between paramters, NS, R2, 95 PPU, r-factor, p-factor, CRPS	Optimal parameter	95% confidence, ARIL	SWAT	Chaohe Basin, China
16	2008	Zhang et al. (2008) ———— SCE, GA, PSO, DE, and AIS		Optimal parameter		SWAT	YRHW, China, RCEW, USA, LREW, USA and MCEW. USA
17	2009	Zhang et al. (2009) ———— GA, and BMA		Optimal parameter	POC (67 and 90%)	SWAT	LREB ,USA and YRHB, China

18	2010	Jiang et al. (2010) PSO, PSSE-PSO and MSSE-PSO		Optimal parameter	X	Xinanjiang model	Tianfumiao reservoir, China
19	2010	Jin et al. (2010) GLUE and MH		Optimal parameter	95% confidence, ARIL	WASMOD	Shuntian catchment, China
20	2010	Jin et al. (2010) GLUE and Formal Bayesian method	Posterior summary statistics for each parameter, Model-fit based on median values of posterior parameter, Average Relative Interval Length (ARIL)	X	95% confidence interval	WASMOD (Xu et al., (1996)	Shuntian catchment, China
21	2010	Keating et al. (2010) NSMC, PEST and DREAM		Optimal parameter	Uncertainty Analysis	Ground Water model	Yucca Flat, Nevada Test Site, USA
22	2010	Li et al. (2010) GLUE and MH		X	95% confidence interval, ARIL,P-95CI, MNS	WASMOD DTVGM	Arid basin, China
	2011	Franz and Hogue (2011) GLUE and SCEM	NSE,Pbias, RMSE, CR, ensemble probability, Brier score		Categorical statistics	SAC-SMA	12 basin ,southern USA
23	2011	Cullmann et al. (2011) PEST, and ROPE	NSE (coefficient of efficiency), RPD (peak flow deviation), RMSE	Optimal parameter	Median, mean and bandwidth of model output based on the ROPE results	WaSiM-ETH (Zappa et al., 2003)	Rietholzbach catchment, Switzerland
24	2011	Jeremiah et al. (2011) AM-MCMC and SMC	Posterior Marginal (Parameter) Distributions, NSE	Optimal parameter	Sensitivity Analysis	Australian Water BalanceModel (AWBM)	Bass River, Victoria
25	2012	Dotto et al. (2012) GLUE, SCEMUA, AMALGAM and MICA	NS, Parameter correlation, Posteriors distributions , POC, ARIL, computational requirements, Number of model runs	Optimal parameter	5% and 95% qunantiles	Urban drainage model KAREN(Rauch and Kinzel 2007)	Richmand, Autrailia

83

26	2012	Kraube and Cullmann (2012) ROPE, AROPE, and ROPE-PSO		95% confidence interval	WaSiM	Rietholzbatch catchment
27	2012	Zhang and Zhan (2012) BNN and MCMC	MSE, CORR, CRPS, % Confidence intervals	X	% Confidence intervals	Neural Networks Model	LREW and RCEW, USA
28	2012	Wang et al. (2012) GA, CGA, and CGASA	RMSE, Peak discharge, time and runoff,	Optimal parameter	X	Xinanjiang model	Shuangpai Reservoir, Chaina
29	2012	Mousavi et al. (2012) SUFI2,PSO and GA	RMSE, hydrograph	Optimal parameter	X	HEC-HMS	Gorganroud River Basin, Iran

Note: X indicates not considered:

5.3   Sampling strategies for uncertainty analysis of hydrological model

Analytical and approximation methods can rarely be applied to complex computer-based models. The probabilistic method of uncertainty analysis is one of the most comprehensive methods used to estimate the uncertainty of hydrological models, which allows the efficient exploitation and improvement of the available physical understanding of complex systems (Montari and Koutsoyiannis, 2012). The Monte Carlo simulation method is a probabilistic method based on random sampling, This is the most common approach to estimate uncertainty of hydrological models. The model outputs associated with a set of inputs and/or parameters are repetitively obtained from a given distribution; subsequently a quantitative estimate of the confidence, which depends critically on sample size, is computed. In low-dimensional problems (i.e., small number of parameters) and simple fast-running models it is possible to sample sufficiently sample many parameter vectors and to cover the parameter space reasonably well (i.e. with high density). In such "data-rich" cases it may be relatively easy to estimate the summary statistics of the resulting distribution (mean, standard deviation etc.). However, in high-dimensional problems and slow-running models the number of sampled points would be lower. Therefore, these models do not cover the entire (highly dimensional) parameter space well. In this case, an "economical" sampling strategy can be used (that still follows the general framework of MC simulation), which allows work with computationally intensive models.

5.3.1 Monte Carlo simulation

The Monte Carlo (MC) method is a flexible and robust method. It is capable of solving a wide variety of problems. In fact, it may be the only method that can estimate the complete probability distribution of the model output for cases with highly nonlinear and/or complex system relationship (Melching, 1995). MC simulation has been widely and successfully applied in hydrological sciences for many years. It helps to estimate model output uncertainty (typically streamflows) resulting from uncertain model parameters, input data or model structure. This approach involves uniform random sampling from the distributions of the uncertain inputs. The model is run successively until a desired statistically significant distribution of outputs is obtained (Beven and Binley, 1992). The main advantages of the MC simulation are that conceptually simplicity, straightforwardness and wide applicability. Nonetheless, this type of method requires a large number of samples (or model runs). The number of samples is often determined by obtaining stable statistics from the distribution of the model output (e.g., Ballio and Guadagnini, 2004, Shrestha et al., 2009).

In MC simulation, random values of each uncertain variable are generated according to their respective probability distributions, and the models are run for each uncertain variable realization. Because we have multiple realizations of outputs from the model, standard statistical technique can be used to estimate the statistical properties (mean, standard deviation, etc.) as well as the empirical probability distribution of the model output.

Consider a deterministic model M of a real-world system predicting a system output variable y given the input data X, the initial condition of the state variables s_0 and the vector of the parameter θ. The model M could be physically based, conceptual, or even data-driven. The model M is also referred to as the "primary model" in order to distinguish it from an uncertainty model that will be described later. For the sake of simplicity, the model M here is referred to here as a conceptual rainfall-runoff model. The system response can be represented by the following equation:

$$y = M(\mathbf{x}, s, \theta) + \varepsilon = \hat{y} + \varepsilon \qquad (5\text{-}1)$$

where ε is the vector of the model errors between the vector of the observed response y and the corresponding model response $\hat{y}$. Note that the state variable s, which appears in the Equation (5-1), will be computed by running the model M given the initial condition of the state variables s_0. Before running the model M, the components of the model (i.e. input data vector $\mathbf{x}$, initial conditions s_0, parameter θ, and the model structure itself) have to be specified, while the output or model response $\hat{y}$ and the state variable s are computed by running the model. These components may be uncertain in various ways to various degrees; the consequences of these uncertainties will be propagated into the model states and the outputs.

To perform the MC simulation, the model is run multiple times by sampling either the input data $\mathbf{x}$ or the parameters vectors θ or even the structure of the model, or a combination of these. Thus, it is mathematically equivalent to the following:

$$\tilde{y} = \tilde{M}(\tilde{\mathbf{x}}, \tilde{\theta}) \qquad (5\text{-}2)$$

where $\tilde{M}$ is the possible candidate of the model structure, $\tilde{x}$ is the sampled input data from the given *pdf*, and $\tilde{\theta}$ is the parameter sampled with the given *pdf* from the feasible domain of the parameter space. For the sake of simplicity the model structure and input data are assumed to be correct. Therefore, Equation (5-2) can be rewritten as:

$$\hat{y}_{t,i} = M(\mathbf{x}, \theta_i); \quad t=1, ..., n; \quad i=1, ..., s \tag{5-3}$$

where θ_i is the set of parameters sampled for the ith run of MC simulation, $\hat{y}_{t,i}$ is the model output of the tth time step for the ith run, n is the number of time steps and s is the number of simulations.

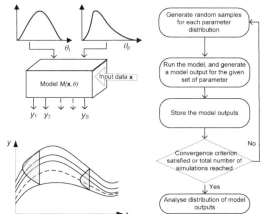

Figure 5-1. Monte Carlo simulation process (Shrestha, 2009)

When MC simulation is used, the error in estimating *pdf* is inversely proportional to the square root of the number of runs s and, therefore, decreases gradually with s. As such, the method is computationally expensive but can reach an arbitrary level of accuracy. The MC simulation is generic, has fewer assumptions, and requires less user-input than do other uncertainty analysis methods. Note that it is difficult to sample the uncertain variables from their joint distribution unless the distribution is well approximated by a multinormal distribution (Kuczera and Parent, 1998). However this problem is however often ignored. The following sampling approaches can be considered when using MC simulation.

a) *Pure random sampling*: all vectors are sampled randomly and independently of each other and the models runs are independent; this approach can be used for low-dimensional cases.

b) *Latin hypercube sampling* (LHS, McKay et al., 1979): the range of each variable is divided into m equally probable intervals, and m sample points are placed at each interval. The number of divisions, m, is equal for each variable. An important feature of LHS is that it does not require more samples for more dimensions (variables), see Figure 5-2.

The LHS technique forces the selection values over the entire parameter range, thereby reducing the total number of samples that requires keeping the probability distribution. Therefore, LHS-based MC simulation is recommended over simple traditional MC simulation.

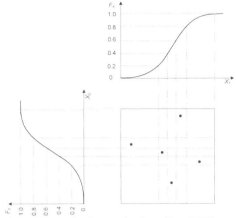

Figure 5-2. Latin Hypercube Sampling (LHS).

c) *Consecutive sampling*: Each consecutive vector depends on the results of the previously generated vectors and on the corresponding model runs. This is an approach taken in the Monte Carlo Markov Chain (MCMC) methods. The most widely used version of this approach is the Metropolis-Hastings (MH) algorithm, a variation of the Metropolis algorithm of Metropolis et al., (1953) that was proposed by Hastings (1970). Other methods, initially developed for randomized search, and by design covering the most important areas of the parameter space, can be used as well, such as the ACCO algorithm (Solomatine, 1999). There are also sampling strategies that combine both approaches, e.g. the SCEM-UA algorithm (Vrugt et al., 2003) (which is a combination of the SCE-UA randomized search algorithm combined with the Metropolis algorithm), or the DREAM algorithm (Vrugt et al., 2009) (which is a combination of differential evolution optimization algorithm with the MCMC scheme).

5.3.2 GLUE

A version of the MC simulation method was introduced under the term "generalized likelihood uncertainty estimation" (GLUE) by Beven and Binley (1992). It uses the concept of behaviour and non-behaviour classification (Spear and Hornberger, 1980). The likelihood measure is used as a criteriation for acceptance or rejection of the models, which is a subjective choice (Freer and Beven, 1996). The behaviour and non-behaviour models are produced by a set of parameters and behaviour models are assigned weights based on likelihood functions (see Equation (5-12).

GLUE is most popular methods that has been widely used over the past 20 years for analyzing predictive uncertainty in hydrological modelling (see, e.g., Freer et al., 1996; Beven and Freer, 2001; Montanari, 2005; Blasone et al., 2008b; Vrught et al., 2008). Users of GLUE are attracted by its easily understandable ideas, relative ease of implementation and use, and its ability to handle different error structures and models without major modifications to the method itself. Despite its popularity, there are theoretical and practical issues related to the GLUE method have been reported in the literature. For example, Mantovan and Todini (2006) argue that GLUE is inconsistent with Bayesian inference processes thereby leading to an overestimation of uncertainty, for both the parameter uncertainty estimation and the predictive uncertainty estimation. A number of authors (Thiemann et al., 2001; Montanari, 2005; Mantovan and Todini, 2006; Stedinger et al., 2008) pointed out that the GLUE does not formally follow the Bayesian approach in estimating the posterior probabilities of parameters and the output distribution. At the same time, Vrugt et al. (2008) presented examples showing that under a variety of different conditions both Bayesian and informal Bayesian methods (like GLUE) can result in similar estimates of predictive uncertainty.

GLUE is sensitive to the selection of the likelihood function and its threshold value (Montanari, 2005, Mantovan and Todini, 2006). The width of the uncertainty bound is dependent on the numbers of behaviour simulations selected for analysis. When large numbers of parameters are used, according the sample sizes are increased (Kuczera and Parent, 1998; Khu and Warner, 2003). This is because the same sampling strategy of MCS is related to computational inefficiency. The spurious correlation among parameters occurs due to sampling procedures used in parameter selection. However, the advancements have been made regarding the efficiency of the GLUE method (e.g., Blasone et al., 2008a; Xiong and O'Connor, 2008). Blasone et al., (2008a) used adaptive MCMC sampling within the GLUE methodology to improve the sampling of the high probability density region of the parameter space. Xiong and O'Connor (2008) modified the GLUE method to improve the prediction limits in enveloping the observed discharge.

5.3.3 MCMC

MCMC generates a sample of parameter values from a constructed Markov chain that converges to the posterior distribution. The chain starts from an initial parameter value, then generates a new value using a proposal distribution, and computes the acceptance probability before determining whether to accept or reject this new value. The variance of the proposal, or candidate, density influences the current state. If this variance is too small, the iterative process will induce a Markov Chain that does not mix rapidly enough over the parameter space. Hence, the tail regions of the posterior may not be sampled sufficiently. If the variance is too large, proposal distribution will frequently be rejected. Convergence of the chain to the posterior distribution is approximated by plotting samples to observe the stability of mean and variance. Gelman and Rubin (1992) proposed a quantitative measurement to check the convergence in both between and within the chain variance.

A numbers of MCMC techniques have been developed for Bayesian sampling of parameter space. The Metropolis Hastings algorithm (Metropolis et al., 1953; Hastings, 1970) is a widely accepted MCMC technique for hydrological modelling (e.g. Kuczera and Parent, 1998; Bates and Campbell, 2001; and Engeland et al., 2005). In the Metropolis methods, the comparison of the statistics of multiple sample chains in parallel provide a formal solution to assess how many model runs are required to reach convergence and obtain stable statistics of the model output and parameters. The sample candidate values can be updated on either each parameter block at a time (block updating) or on each parameter at a time (single-size updating). We use single-size updating of the Metropolis-Hastings algorithm (Marshall et al., 2004) in this thesis.

Oftentimes an inappropriate selection of the proposal distribution in the Markov Chain causes slow convergence. The efficiency can be improved by using an adaptive evolutionary learning strategy within MCMC sampling. An example of such methods are delayed rejection adaptive Metropolis (Haario et al., 2006) and differential evolution adaptive Metropolis (Vrugt et al., 2008).

5.3.4 SCEMUA

MCMC sampler entitled the shuffled complex evolution metropolis algorithm (SCEMUA) was presented by Vrugt et al. (2003). This algorithm is a modified version of the SCE-UA global optimization algorithm developed by Duan et al. (1992). The SCEMUA operates by merging the strengths of the Metropolis algorithm, controlled random search, competitive evolution, and shuffled complex to continuously update the proposal distribution and evolve the sampler to the posterior target distribution. In contrast with traditional MCMC samplers, the SCEMUA algorithm is an adaptive sampler, the covariance of the proposal distribution is periodically updated in each complex during the evolution to the posterior target distribution using information from the sampling in the transitions of the generated sequence. The SCEMUA algorithm has been successfully applied in hydrological application. For example, it has been utilized in the conceptual rainfall–runoff model HYMOD (Vrugt et al., 2003), the groundwater model MODHMS (Vrugt et al., 2004), LISFLOOD model (Feyen et al., 2007; 2008), the rainfall–runoff model TONET (McMillan and Clark., 2009), and in artificial neural network (ANN) rainfall-runoff model (Guo et al., 2012).

5.3.5 DREAM

Vrugt et al. (2008) demonstrated a significant improvement in the efficiency of MCMC sampling by using a self-adaptive differential evolution learning strategy in population-based sampling. This approach is known as differential evolution adaptive Metropolis (DREAM) algorithm. This algorithm is a modified version of the SCEMUA but reportedly has the advantage of maintaining detailed balance and ergodicity while presenting good efficiency for complex, highly nonlinear and multimodal target distributions. The DREAM runs multiple different chains simultaneously for global exploration. It also automatically tunes the scale and orientation of the proposal distribution in randomized subspaces during the search. The method starts with an initial population of points (parameter sets) to strategically sample the parameter space of potential solutions. Ergodicity of this algorithm showed that this algorithm is superior among other adaptive MCMC sampling approaches (Vrugt et al., 2009). In addition, this algorithm significantly enhances the applicability of MCMC simulation to complex, multi-modal search problems.

Recently, Laloy et al. (2010) implemented the calibration of continuous, spatially distributed, process-based and plot-scale runoff model. Keating et al. (2010) compared null-space Monte Carlo with DREAM in the optimization and uncertainty analysis of ground water parameters. Minasny et. al. (2011) explored the prediction of the uncertainty interval of the posterior distribution of geo-statistical parameters using DREAM algorithm. He et al. (2011) investigated the sensitivity and uncertainty of snow parameters combining with DREAM with generalize sensitivity analysis (Hornberger and Spear, 1981). Laloy and Vrugt (2012) solved the search problems of high-dimensional (i. e., 241 number of parameters) parameters in the spatially distributed hydrologic model using the modified DREAM in a distributed computing environment.

5.3.6   ACCO

The adaptive cluster covering algorithm (ACCO) was developed by Solomatine (1999). This is a random search global optimization algorithm based on the principles of clustering, covering, adaptation and randomization. These principles are briefly described in Section 2.4.1 and a detailed description of the algorithm can be found in Solomatine, (1999). ACCO has been successfully applied for the parameter optimization of hydrological models (e. g., Shrestha et al., 2009; Kayastha et al., 2013).

5.3.7   PSO

The Particle Swarm Optimization (PSO) algorithm has received significant amount of attention from researchers due to its simplicity and promising optimization ability in various problems. This optimization algorithm was developed by Kennedy and Eberhart (1995) to solve the problem of the social behaviour of bird flocking or fish schooling. The algorithm uses a group-based stochastic optimization technique for continuous nonlinear functions. It is characterized by a simple concept adapted from the decentralized and self-organized systems found in nature , whereby all of the particles move to get better results. The PSO algorithm runs primarily with a group of random particles assigned with random positions and velocities. Through a series of iterations the algorithm searches for optima as the particles are escalated through the hyperspace searching for possible solutions. PSO has been successfully applied for the parameter estimation of hydrological models (e.g., Gill et al., 2006; Zhang et al., 2008; Chu and Chang, 2009; Chou, 2012).

Jiang et al. (2010) pointed out the presence of an early convergence problem in PSO. A particle in the swarm can find its current optimal position. However, if this position is in a local optimum, then the particle swarm will not be able to search over again in the solution space. As a result the algorithm traps into local optima. Shi et al. (2005) presented a hybrid evolutionary algorithm based on PSO and GA methods, which possess better ability to find the global optimum than does the standard PSO algorithm.

5.4   Characterization of uncertainty

5.4.1   Prediction interval

The problems of hydrology involve modelling of complex hydrological processes over a space of time, which constitute in time series of observations and outputs of the model. The model is composed of the set of equations that represent rainfall runoff relation. For example, see Equation (5-1), The uncertainty associated with the parameters is considered and MC simulation is performed by running the model M multiple times by changing the parameter vectors. We assume that the model structure and input data are correct and the parameter is sampled with the given *pdf* from the feasible domain of the parameter space. This is expressed by Equation (5-3).

The statistical properties (e.g., quantiles) of the model output for each time step t are estimated from the realizations $\hat{y}_{t,i}$. The uncertainty of model output is usually described by informative quantities such as prediction intervals and quantiles. Prediction interval is between upper and lower limits, wherein a future unknown value with prescribed probability is expected to lie. These limits are typically the quantiles of the model output distribution. In each simulation, model output is given a different weight as in the case of GLUE. The transferred prediction quantile $Q(p)$ corresponding to the pth [0, 1] quantile can be calculated by the formula:

$$P(y_t < Q_t(p)) = \sum_{i=1}^{s} w_i |y_{t,i} < Q_t(p)| \qquad (5\text{-}4)$$

where w_i is the weight given to the model output at simulation i, y_t is the realization vector at time step t, and $y_{t,i}$ is the value of model outputs at the time t simulated by the model $M(x,\theta_i)$. At simulation i, $Q(p)$ is the $p\%$ quantile. The prediction interval $[PI_t(\alpha)]$ is derived from the transferred prediction quantile for the given confidence level of $1-\alpha$ $(0<\alpha<1)$

$$\begin{aligned} Q_t(p) &= PL_t^L, & where, \quad p &= (1-\alpha)/2 \\ Q(p) &= PL_t^U, & where, \quad p &= (1+\alpha)/2 \\ PI_t &= PL_t^U - PL_t^L \end{aligned} \qquad (5\text{-}5)$$

where, PI_t is the distance between the upper PL_t^U and lower PL_t^L prediction limits and refers to the prediction interval corresponding to the $1-\alpha$ confidence level.

5.4.2   Uncertainty indices

The following different uncertainty indices can be used to evaluate the uncertainty in the model output prediction:

1. Prediction interval coverage probability (*PICP*): It measures the percentage of the number of observations enveloped by the prediction intervals. The higher this percentage value, the better in terms of representing the uncertainties.

$$PICP = \frac{1}{N} \sum_{i=1}^{N} C_t \qquad (5\text{-}6)$$

$$with \quad C = \begin{cases} 1, PL_t^L \le y_{t=1} \le PL_t^U \\ 0, otherwise \end{cases}$$

where y_t is the observed model output at the time t.

2. Mean prediction interval (*MPI*): This is the average width of the uncertainty region (prediction intervals). This metric expresses the width of the uncertainty of the model simulation is as well as the narrowness of the confidence intervals. This is calculated using the formula below :

$$MPI = \frac{1}{N} \sum_{i=1}^{N} \left(PL_t^U - PL_t^L \right) \qquad (5\text{-}7)$$

3. Average asymmetry degree (*S* index and *T* index): These are the indices for assessing the geometric structure of the band formed by prediction intervals. They calculate the degree of asymmetry of the prediction intervals with respect to observations:

$$S = \frac{1}{N}\sum_{i=1}^{N}\left|\frac{PL_t^U - y_t}{PL_t^U - PL_t^L} - 0.5\right| \tag{5-8}$$

$$T = \frac{1}{N}\sum_{i=1}^{N}\left(\frac{\left|\left(PL_t^U - y_t\right)^3 + \left(PL_t^L - y_t\right)^3\right|}{\left(PL_t^U - PL_t^L\right)^3}\right)^3 \tag{5-9}$$

4. Average deviation amplitude and relative deviation amplitude (D and RD): These indices demonstrate the deviation of the middle point of the prediction intervals from the corresponding observations.

$$D = \frac{1}{N}\sum_{i=1}^{N}\left|\frac{1}{2}\left(PL_t^U + PL_t^L\right) - y_t\right| \tag{5-10}$$

$$RD = \frac{1}{N}\sum_{i=1}^{N}\left(\left|\frac{1}{2}\left(PL_t^U + PL_t^L\right) - y_t\right|/y_t\right) \tag{5-11}$$

where y_t is the observed model output at the time t. Together with these uncertainty indices, a visual inspection of the plot could provide additional significant information along the different regimes (e.g., high flows) that are produced by different algorithms. The above motioned uncertainty indices given in Equation (5-6) and Equation (5-7) can be found in Shrestha et al. (2009) while those indices given in Equations (5-8) to (5-11) can be found in Xiong et al. (2009).

5.4.3 Likelihood functions

The notion of likelihood is used to estimate the model performances. Classical formulation defines likelihood as the probability of observing the sampled data set if a certain sampling distribution is used. The likelihood function is used to measure the degrees of belief in model prediction indicating how well the model can reproduce observation. An inappropriate selection of the likelihood function can lead to an incorrect quantification of the model uncertainty. Generally, the likelihood is calculated from the probability of the model output that generated by a set of parameter values. A likelihood is categorized into two types; formal Bayesian and informal likelihoods. A number of studies exist for definitions (e.g., Vrugt et al, 2008; Stedinger et al., 2008; McMillan and Clark, 2009; Scoup and Vrught, 2010). In the present study, we use the following two likelihood functions:

5.4.3.1. Informal likelihood

This likelihood is normally calculated based on model error variance. The sum of the squared errors is the basis to calculate the (generalized) likelihood measure (see, Freer et al., 1996). The formula is stated as follows:

$$L(\theta_i / y) = (1 - \frac{\sigma_e^2}{\sigma_{obs}^2})^\lambda \qquad (5-12)$$

where, $L(\theta_i/y)$ is the likelihood measure for the ith model (with parameter vector θ_i) conditioned on the observations y, σ_e^2 is the associated error variance for the sth model, σ_{obs}^2 is the observed variance for the period under consideration, and λ is a user defined parameter. When λ set to 1, the Equation 5-12 returns the equivalent to Nash-Sutcliffe model efficiency (NSE), which is one of the most widely used performance measures in hydrology. This is calculated as 1.0 minus the normalization of the mean squared error by the variance of the observed data; its value varies between minus infinity to 1.0 (Nash and Sutcliffe, 1970).

5.4.3.2.   Formal Bayesian likelihood

The probability density function (pdf) of the residual errors is specified a priori. This prior information about parameters usually consists of lower and upper ranges for each of the parameters as feasible parameter space. It assumes a non-informative uniform prior distribution, and the residuals are mutually independent, each having the exponential power density $E(\sigma, \gamma)$ and the likelihood of a parameter set θ for describing the observed data y. All these forms can be derived using the Box and Tiao (1973) equation.

$$L(\theta_i \mid y, \gamma) = \left[\frac{\omega(\gamma)}{\sigma}\right]^N \exp\left[-c(\gamma)\sum_{s=1}^{N}\left(\frac{e_s(\theta_i)}{\sigma}\right)^{2/(1+\gamma)}\right] \qquad (5-13)$$

where θ is the given parameter set, N is the length of the data series, and $e_s(\theta_i)$ denotes error at sth observed data series when given the parameter set θ. If we assumed (following Vrugt et Al. 2003) that the influence of σ is integrated out by assuming a non-informative prior of the form $p(\theta, \sigma | \gamma) \propto 1/\sigma)$ then the likelihood function would be written as:

$$L(\theta_i \mid y, \gamma) = C[M(\theta_i)]^{-N(1+\gamma)/2} \qquad (5-14)$$

C is normalized integral, $C^{-1} = \int_\Theta [M(\theta_i)]^{-N(1-\gamma)} d\theta_i$, and θ is the feasible space of the parameter:

$$M(\theta_i) = \sum_{s=1}^{N} |e_s(\theta_i)|^{-2(1+\gamma)} \qquad (5-15)$$

For convenience, we use the log-likelihood function as follows:

$$L(\theta_i | y, \gamma) = \log C - \frac{N(1+\gamma)}{2}\log[M(\theta_i)] \qquad (5-16)$$

5.5   Experiment setup for the Nzoia catchment

The Nzoia catchment is considered as the case study (see Chapter 1 for a description of the catchment). The HBV model was set up with nine model parameters for calibration. The data period was from 1-Jan-1970 to 31-Dec-1979, comprising 3544 daily data. The first of two months of data were considered as model warm up period. Several sampling-based algorithms were run to estimate the uncertainty of the hydrological model. Parameters were sampled using non-informative uniform random sampling (without prior knowledge of individual parameter distribution, but selected a feasible range of values which are presented in Table 5-2). The choice of prior distribution of parameters is subjective and feasible ranges were determined by the physical meaning of parameters with the model behaviour (Pappenberger et al., 2007). The three experiments were setup based on feasible ranges of parameters, which are described as follows:

1. Experiment 1 (EX1): For the first experiment, the model runs with a different combinations of parameters (sampling of parameters) chosen from the ranges of parameters set up based on the calibrated values from the other model applications and information about the catchment. These are presented in Table 5-2 as Type-I range and widths (white and grey), as shown in Figure 6-3.

2. Experiment 2 (EX2): In this experiment, model uncertainty was analysed with exclusively behavioural models. These models were selected from the samples generated by the first experiment. The output generated by each set of parameters was assigned a likelihood value, which represents the capability of simulating observed responses. The total sample of simulations was divided into behavioural and non-behavioural based on a cut-off threshold above the value of the likelihood value (*NSE*). Negative *NSE* values were rejected for further analysis of uncertainty.

3. Experiment 3 (EX3): In this experiment, the model runs with parameters that were sampled from Type II range of parameters. The Type II were drawn from Type I after optimum parameters found in the first experiment and these are presented in grey in Figure 5-3.

Figure 5-3. Ranges of parameters: (a) white bar is ranges Type-I and (b) grey bar is rangesType-II (narrow)

Table 5-2. Ranges of HBV model parameters

Descriptions and unit of parameters	Ranges	
	Type I	Type II
FC (Maximum soil moisture content), - (mm)	50 - 600	500 - 600
LP (Limit for potential evapotranspiration),	0.1 - 1	0.1 - 1
ALFA (Response box parameter) ,	0 - 4	0 - 2
BETA (Exponential parameter in soil routine),	1- 6	1 - 2
K (Recession coefficient for upper tank), (/day)	0.05 - 0.5	0.05 – 0.5
K4 (Recession coefficient for lower tank), (/day)	0.01 - 0. 3	0.01-0.1
PERC (Maximum flow from upper to lower tank), (mm/day)	0 - 8	5 - 8
CFLUX (Maximum value of capillary flow), (mm/day)	0 - 1	0 - 0.5
MAXBAS (Transfer function parameter), (day)	1 – 3	2 - 3

MC simulation convergence analysis

MC simulations run for a fixed number of times. The model runs for each random parameter set, and the objective function is computed for each model run. *NSE* is used as the basis to calculate objective function. The adequate number of samples can be confirmed by the convergence of variation in the average model error across all model runs. This is set by comparing two statistics, mean and standard deviation of the objective function (*NSE*)

$$ME_k = \frac{1}{k}\sum_{i=1}^{k} NSE_i \qquad\qquad (5\text{-}17)$$

$$SDE_k = \sqrt{\frac{1}{k}\sum_{i=1}^{k}(NSE_i - ME_k)^2} \qquad\qquad (5\text{-}18)$$

where, NSE_i is the coefficient of model efficiency of the *i*th MC run, and ME_k and SDE_k are the mean and standard deviation of the model efficiency up to the *k*th run, respectively.

In the first experiment, MC simulations generated 500,000 samples. This was done in order to obtain reliable results and to ensure that random samples adequately covered the complete range of parameters. The convergence analysis showed that both statistics stabilized after 10,000 simulations. However 25,000 simulations are considered reasonable for further analysis (corresponding model outputs were used to calculate the prediction interval).

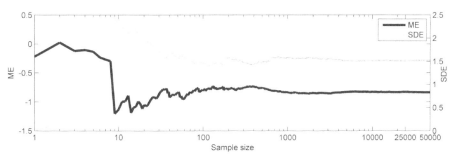

Figure 5-4: The convergence of the mean (*ME*) and the standard deviation (*SDE*) of Nash-Sutcliffe model efficiency (*NSE*).

The GLUE method uses the rejection of the subjective value of *NSE* during sampling. In experiment EX1, the number of behavioural samples retained 165330 out of 500000 MC samples using rejection threshold value of 0. However, we used only 25000 samples for further evaluation and the behavioural sample retained 8260 samples. According to the percentage of behavioural samples corresponding to the rejection threshold as measured by *NSE* observed that only approximately 33% of simulations were accepted. In contrast, more than 82% of simulations were retained with the sample produced from a narrow range of parameters (in Experiment 3). The corresponding model outputs based on behavioural samples were used to build the prediction interval.

Using the MCMC approach, arbitrary values were first sampled from the selected parameter space using a uniform distribution as a proposal distribution with five parallel chains. The Metropolis Hastings (MH) algorithm (Hastings, 1970), was applied to estimating the posterior parameter distributions and likelihood was calculated using Box and Tiao (1973) (see, Equation (5-13)). Each chain of 5,000 samples is retained with an acceptance rate ranging between 40% and 50%. The scale reduction score (Gelman and Rubin, 1992) was performed to check the convergence of the sampler to stationary point based on sequence variances. Its formula is written as follows:

(5-19)

$$SDS = \sqrt{\frac{n-1}{n} + \frac{q+1}{q*n} * \frac{B}{W}}$$

where, $\quad B = \frac{n}{q-1}\sum_{j=1}^{q}\left(\theta_j - \overline{\theta}\right)^2 ; \quad \overline{\theta} = \frac{1}{q}\sum_{j=1}^{q}\theta_j ; \quad \theta_j = \frac{1}{n}\sum_{i=1}^{n}\theta_{ij}$

$$W = \frac{1}{q}\sum_{j=1}^{q}s_j^2; \quad\quad s_j^2 = \frac{1}{n-1}\sum_{i=1}^{n}\left(\theta_{ij} - \overline{\theta_j}\right)$$

where θ_{ij} is parameter (i=1,.., n; j=1,.., q), n is the number of simulations in each sequence q, W is the variance between q means, and B is the average of q within the sequence for parameter. The scale reduction score of 1.2 was considered for convergence of the Markov chain. After approximately 5,500 simulations, there was convergence to the stationary posterior distribution of each parameter. A total of 25,000 samples (q=5 with each of n=5000) were selected for an uncertainty evaluation.

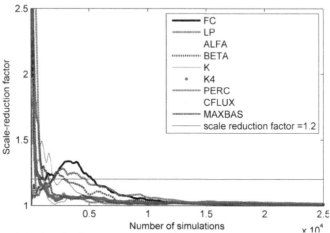

Figure 5-5. Scale reduction score generated by MCMC for parameters of HBV

The SCEM-UA algorithm runs with 250 initial populations of parameters and 10 complexes. The parallel sequence from each complex was partitioned from the population, the likelihood value of each parameter set is calculated by formal Bayesian likelihood (see Equation (5-13)). The complete set of 25000 samples were included which was a large enough sample from the high probability density region. The Gelman-Rubin method was used for the convergence diagnostics of each parameter as well as random initializations of the starting points of each of the parallel sequences. The scale reduction factor was quite large for the first 10,000 samples generated. Thereafter, the convergence diagnostics for each of the parameters narrowed down very quickly.

A total of 25080 model evaluations with 9 Markov chains were performed over the LHS-specified samples using DREAM. The same values used to define the upper and lower bounds for the other methods were assigned here as well (Table 5-2). First the model ran until reaching the maximum number of evaluations; however 3,500 samples were sufficient to find the best parameters, indicating that this algorithm has a faster convergence campared to other MCMC samplers. Furthermore, the number of individual chains with different starting points helped in dealing with multiple regions of highest attraction and facilitating the powerful array, and leading to a fast convergence.

For the PSO algorithm the total number of particle swarms, maximum functional evolutions, and the cognitive and social acceleration coefficients of 500, 25000, 2.1, and 2.1, respectively were used to run the model. The model started with a set of randomly generated parameters and updated the swarm in each iteration. This process continued until the stopping criteria were satisfied. If we used a small particle swarm size, then the algorithm would stop before the stopping criteria were satisfied. Therefore, for this study, we selected a 500 particle swarm size to obtain the maximum functional evaluation, and then estimated the uncertainty of model outputs.

ACCO is by design a randomized search algorithm that typically requires few function evaluations. A total of 18,600 samples were used, which is a large enough to find the best set of parameter among the given parameter ranges. An initial population of 1000 and cluster population of 500 were used to run ACCO. Consequently, it generated a smaller number of points compared to other algorithms. Indeed the ACCO was the fastest at finding the best sets of parameters; however, a reasonable number of initial sets and for clustering should be provided. All of the algorithms mentioned require adequate parameterizations in order to ensure that sampled parameters cover the selected ranges. The best value set of parameters and *NSE* is presented in Table 5-1.

5.6 Experimental results and discussion

5.6.1 Distribution of the model objective function

The objective function (*NSE*) was assigned to the respective model output generated from model runs with associated parameter sets in each algorithm. Functional evaluations (i.e., number of simulations) of the model outputs resulting from each algorithm are shown in Figure 5-6. These are categorized into three groups: (i) highly varied objective function over completely functional evaluations, such as MCS and GLUE, (ii) less deviated of the objective function, such as MCMC, (iii) simultaneously narrows down, such as SCEMUA, DREAM, PSO, and ACCO. The first and second groups are flat in distribution compared with third group because the objective functions are calculated by randomly generated parameters from a large number of points exploring the entire parameter space. The third group algorithms are designed to search for the best parameters of the model; therefore typically moves toward the region of low objective function value can be achieved. The functional evaluations are widespread in the beginning because the initial population was used for better search of space and later functional evaluations decreased to a lower value of the objective function.

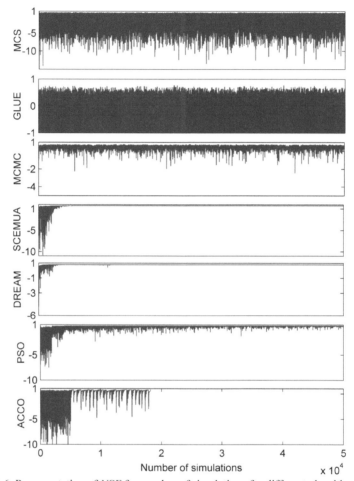

Figure 5-6. Representation of *NSE* for number of simulations for different algorithm (EX1)

The shape of the cumulative probability distributions (*CDFs*) are used to determine the deviation of model evaluations (*NSE*). These deviations are evaluated by model outputs from different sampling algorithms. The *CDFs* of *NSE* by different algorithms are shown in Figure 5-6 and Figure 5-7, where different numbers of samples were used in each case. The slope of *CDFs* revealed the spread of the distribution of *NSE*. Both MCS and GLUE showed wide distributions, followed by MCMC. SCEMUA and DREAM had steep gradients. MCS, GLUE and ACCO had monotonous gradients toward the higher value of the objective function. It should be noted that the shape of the distribution of *NSE* makes a significant contribution to the narrow context and wide uncertainty bound. The monotonous slop ensemble contains a wider uncertainty range and the steep gradient produces narrow uncertainty. Large numbers of *NSE* values are obtained below zero value when sampling parameters from a wide range of parameters (EX1). However, fewer portions of negative *NSE* values remained when sampling a narrow range of parameters (EX3).

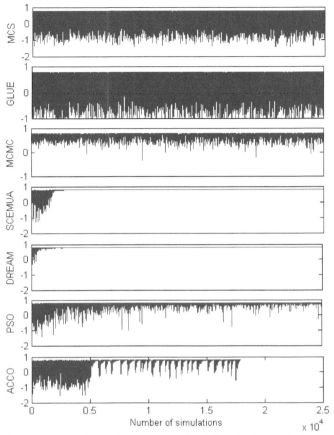

Figure 5-7. Representation of *NSE* for number of simulation for different algorithm (EX3)

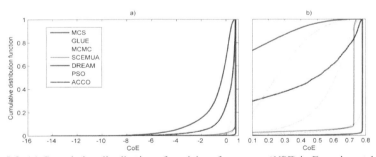

Figure 5-8. (a) Cumulative distribution of model performances (*NSE*) in Experiment 1 and

(b) fragment from (a))

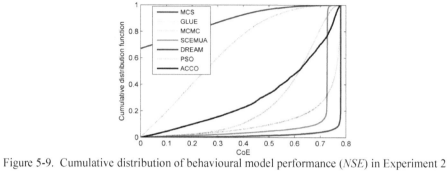

Figure 5-9. Cumulative distribution of behavioural model performance (*NSE*) in Experiment 2

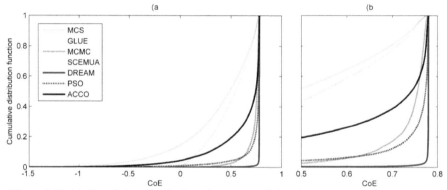

Figure 5-10. (a) Cumulative distribution of narrow model performance (*NSE*) in Experiment 3 and (b) fragment from (a)

5.6.2   Parameter posterior distribution

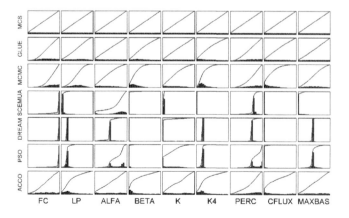

Figure 5-11. Posterior distribution of parameters (Experiment 1)

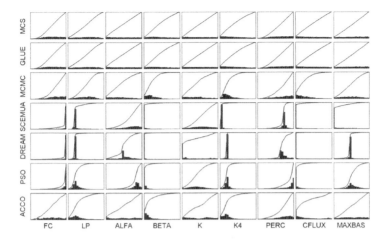

Figure 5-12. Posterior distribution of parameters (Experiment 2)

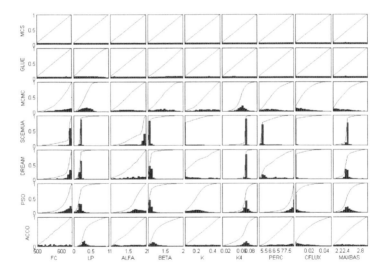

Figure 5-13. Posterior distribution of parameters (Experiment 3)

The probability will be higher for the bins with many points and lower error, while the probability will be lower for the bins with the higher error. The best parameter sets from all algorithms showed almost equal performance (*NSE*). However, they are resulted from different numbers of evolutions and their parameter values differ. These different parameter sets represent equivalent performance, highlighting the equifinality concept (Beven and Binley 1992). More time was required for searching the best parameter sets in Type –I range of parameters. If the parameter range (Type-III) was narrow, the bounds obtained for both the distribution of posterior parameters and model outputs would be narrower. However, if the uncertainty bounds were too narrow, the model could be unacceptable to compare with the observed value.

Table 5-3. Best parameter sets and performances (Experiment 1)

Parameters	MCS	GLUE	MCMC	SCEMUA	DREAM	PSO	ACCO
FC	566.841	566.841	538.230	585.977	594.875	594.800	589.427
LP	0.255	0.255	0.307	0.381	0.283	0.283	0.299
ALFA	3.079	3.079	1.458	1.986	1.922	1.913	3.941
BETA	1.071	1.071	1.176	1.244	1.094	1.094	1.096
K	0.360	0.360	0.436	0.147	0.050	0.050	0.385
K4	0.076	0.076	0.070	0.049	0.070	0.070	0.062
PERC	6.373	6.373	5.283	7.390	5.344	5.342	6.220
CFLUX	0.021	0.021	0.005	0.052	7.799E-09	4.8E-10	0.002
MAXBAS	2.711	2.711	2.081	2.071	2.404	2.403	1.922
Value of NSE	0.7733	0.7733	0.7751	0.7599	0.7801	0.7801	0.778
Number of function evaluations	500,000	165,330	100,000	100,000	100,000	100,000	18,200
Number of function evaluations used for uncertainty evaluation	25,000	8,260	25,000	25,000	25,000	25,000	18,200

Table 5-4. Behavioural samples by threshold value of 0 in NSE (Experiment 2)

	MCS	GLUE	MCMC	SCEMUA	DREAM	PSO	ACCO
Number of function evaluations	25,000	8,260	24,603	23,917	24,886	23,916	13,403

Table 5-5. Best parameter vectors and performances (Experiment 3)

Parameter	MCS	GLUE	MCMC	SCEMA	DREAM	PSO	ACCO
FC	531.99	531.99	597.69	594.87	593.23	592.89	582.53
LP	0.263	0.263	0.235	0.283	0.285	0.283	0.288
ALFA	1.268	1.268	1.441	1.920	1.799	1.726	1.633
BETA	1.120	1.120	1.024	1.094	1.099	1.097	1.116
K	0.389	0.389	0.323	0.050	0.051	0.152	0.408
K4	0.069	0.069	0.066	0.070	0.070	0.070	0.071
PERC	6.698	6.698	7.953	5.339	5.319	7.910	7.633
CFLUX	0.004	0.004	0.001	7.619E-09	2.513E-07	7.384E-06	3.118E-05
MAXBAS	2.410	2.410	2.675	2.403	2.411	2.422	2.474
Value of NSE	0.7769	0.7769	0.7796	0.7801	0.7801	0.7801	0.78
Number of function evaluations	25000	21381	25000	25000	25000	25000	17929

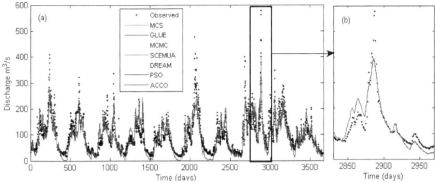

Figure 5-14. Hydrographs of observed and best-simulated discharges from various sampling methods
(01/01/1970 -31/12/1979)

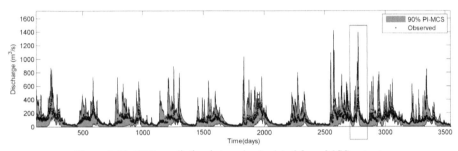

Figure 5-15. 90% prediction interval calculated from MCS outputs

Due to the limits of computer memory, it is problematic to compute the prediction intervals for an ensemble of many model outputs in one time. One solution to this is to use the block wise calculation method, where the prediction intervals are calculated part by part. For this analysis, the generated model outputs are partly saved in a vertical block and then the saved vertical blocks are loaded one by one and split horizontally into a small blocks. Each horizontal small block is merged vertically and then the prediction interval is calculated for that block. This process is repeated until the final prediction intervals are obtained. For example, to calculate the prediction interval of 25,000 model realizations with 3,500 of data points (time series), the blocks are first split and saved as 50 blocks each containing 500 model realizations. These blocks are loaded again, one by one and each block split into 10 parts. Each block size would bc 500 by 350. At this points, the blocks are merged vertically (making the block size 2,5000 by 350). The prediction interval for that block is then calculated (i. e., prediction interval for 350 data points). The blocks are loaded repetitively and the prediction intervals are calculated until the full length is obtained (3,500 data points). This type of block-wise analysis solves the memory problem that arises from working with large data (uncertainty calculation from model outputs) on a single computer. It can be used for any kind of large data.

103

5.6.3 Statistical analysis of results

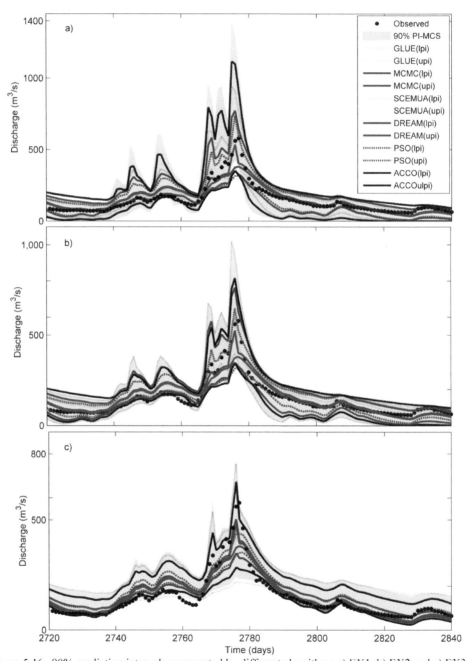

Figure 5-16. 90% prediction intervals represented by different algorithms a) EX1, b) EX2 and c) EX3

The last row of Table 5-3 lists the numbers of function evaluations from each algorithm. However, the GLUE and ACCO used less than 25,000 samples for uncertainty evaluations.

GLUE used the accepted samples (non-negative *NSE*). The ACCO generated a smaller number of points and stopped when one best point was found ignoring the areas with the lower *NSE*. The uncertainty bounds were calculated as the 90% prediction intervals from the generated model outputs, that is at each time step of the simulation, the output estimated 5% and 95% quantiles of the distribution. If these bounds were too wide, then there was sufficient space to include most of the observations.

Figure 5-16 illustrates important differences among the uncertainty bounds calculated from the model outputs that were generated from different algorithms. The uncertainty bound was estimated based-on the probability distribution of model outputs generated by parameters sampled from a given distribution. The 90% prediction intervals represented the uncertainty bound that should be able to capture the all observed streamflows, but in reality may not due to no representation of all sources of uncertainty. We considered only parameter uncertainty with the assumption that the model structure, input data (such as rainfall, temperature data), and output discharge data are correct. The wide uncertainty bounds were observed from MCS, GLUE and ACCO in all experiments. This may be due to the fact these three methods explore the parameter space more thoroughly than the others do. As a result, relatively more bad models (with high error and low likelihood) may have come in when we calculate the quantiles of the discharge distribution leading to a wider prediction interval. The DREAM produced narrow uncertainty bands in its predictions

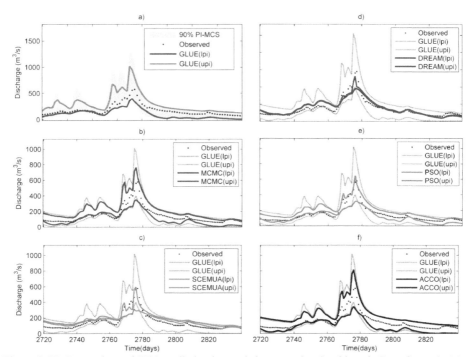

Figure 5-17. Comparison of 90% prediction intervals between the algorithms in Experiment 2: (a) MCS–GLUE, (b) GLUE–MCMC, (c) GLUE–SCEMUA, (d) GLUE–DREAM, (e) GLUE–PSO, (f). GLUE–ACCO.

Table 5-6. Uncertainty indices

Experiment 1

Indices	MCS	GLUE	MCMC	SCEMUA	DREAM	PSO	ACCO
PICP(%)	80.77	78.65	67.82	13.03	9.73	72.22	86.60
MPI	172.00	103.57	74.30	10.70	7.27	67.95	133.99
S	0.41	0.38	0.55	3.07	4.60	0.46	0.27
T	0.88	0.84	1.04	4.05	5.94	0.94	0.72
D	37.63	25.31	24.33	24.34	22.65	22.06	27.10
RD	0.44	0.31	0.32	0.30	0.31	0.29	0.31

Experiment 2

PICP(%)	80.77	78.65	66.53	20.78	8.96	67.19	80.14
MPI	172.00	103.57	72.15	17.83	6.61	57.34	88.92
S	0.41	0.38	0.57	1.60	5.06	0.51	0.31
T	0.88	0.84	1.06	2.25	6.52	1.00	0.77
D	37.64	25.91	24.38	25.11	22.58	21.64	23.28
RD	0.44	0.31	0.32	0.31	0.31	0.28	0.28

Experiment 3

PICP (%)	93.97	93.69	36.60	9.93	9.56	52.38	85.22
MPI	117.78	114.11	25.72	6.35	7.08	40.66	86.68
S	0.23	0.22	1.17	4.94	4.73	0.77	0.28
T	0.68	0.67	1.76	6.36	6.11	1.30	0.74
D	27.14	25.08	21.90	22.26	22.33	21.98	23.59
RD	0.34	0.31	0.28	0.30	0.30	0.29	0.28

Figure 5-17 depicts the comparison of GLUE and the other algorithms based on uncertainty bounds that were estimated by the Type I parameter ranges. The GLUE and ACCO are capable of covering the observation at peak streamflows; however, the PSO, SCEMUA, and DREAM did not cover the observation. Using the SCEMUA and DREAM, most of the peaks and other limbs stayed outside the prediction intervals.

The accuracy of uncertainty was measured by the uncertainty indices- *PICP, MPI, S, T, D* and *RD* which are presented in Table 5-6 (Their formulas are presented in Equations (5-6) to (5-11). All of the indices except *MPI* were evaluated based on the observed streamflows. In all these indices, the experiments revealed a high correlation between the width of the uncertainty bounds and the ranges of parameters. *PICP* was expected to be perfect when its value becomes higher (perfect value is 100%). The results of the Experiment 3 showed that the higher percentage of (93.97% and 93.69 %) of observed streamflows were covered by the MCS and GLUE respectively in narrow ranges of parameters. However, smaller values (36.60%, 9.93% and 9.56 %) were covered by MCMC, SCEMUA, DREAM and PSO. DREAM and ACCO do not greatly alter the value of *PICP* with respect to parameter ranges (Type I and Type II). The ACCO produced a lower *MPI* than did MCS in wide ranges of parameters, although *PICP* was higher. The GLUE produced a high *MPI* when parameter ranges were narrow, as it rejects very few non-behavioural models, although the width of the uncertainty bound varied with the rejection threshold. When a narrow range of parameters was used, the values of *S* and *T* were increased for the MCMC, SCEMUA, DREAM, PSO and ACCO. The values of *D* and *RD* decreased for the ACCO with leftover values retaining only small behavioural parameters changes.

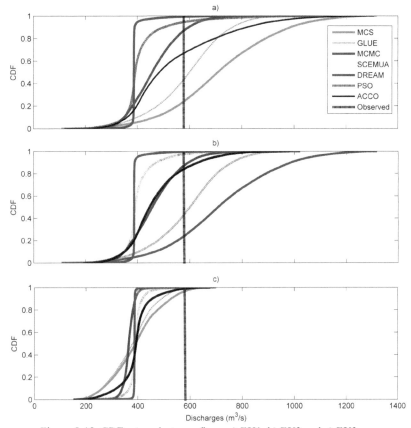

Figure 5-18. CDFs at peak streamflows a) EX1, b) EX2 and c) EX3

Figure 5-18 illustrates the relative width of the uncertainties in the predictions at peak streamflows. The observed streamflows reasonably covered the upper and lower distributions of streamflows at peak level in the MCS, GLUE and ACCO, while MCMC SCEMUA, DREAM, and PSO covered significantly less in EX 1 and EX 2. However, in EX 3 and did not cover the peak at all. The sharp gradient in the distribution represented narrow uncertainty bounds in their predictions.

This study also tests the CDFs and uses Kolmogorov-Smirnov analysis in order to compare the significant differences of two datasets at peak streamflows.

Table 5-7. K-statistics from Kolmogorov-Smirnov test

		GLUE	MCMC	SCEMUA	DREAM	PSO	ACCO
EX1	MCS	0.31	0.63	0.85	0.90	0.77	0.44
EX2	MCS	0.31	0.63	0.89	0.90	0.80	0.61
EX3	MCS	0.07	0.41	0.42	0.47	0.31	0.18

The Kolmogorov-Smirnov test was carried out to compare two datasets (one from MCS and another from selected algorithm) in order to evaluate their significant differences. These results are shown in Table 5-3. The values obtained by GLUE, MCMC and DREAM in EX 1 and EX2 are similar, while the others are not. The values obtained using the DREAM and SCEMUA are quite different from those obtained using the GLUE.

5.7 Summary

This chapter presents a comparison of seven different sampling-based uncertainty methods for hydrological models. The uncertainty of model outcomes from the various sampling algorithms are estimated as two quantiles of the *pdf*, and these are analyzed by different uncertainty indices. The numbers of good data-fitting regions are identified in the parameter ranges by plotting the posterior distribution of the parameters. GLUE method requires subjective decisions in the selection of a cutoff threshold that separates behavioral from non-behavioral parameter sets. MCMC simulation is a widely adopted approach to estimate the posterior probability distribution function of the parameters, and it appropriately samples the high-probability-density region of the parameter space. SCEMUA is a global optimization algorithm that provides an efficient estimate of the most likely parameter set as well as its underlying posterior probability distribution within a single optimization run. In the DREAM method, the model runs multiple chains simultaneously for global exploration of parameter space and automatically tunes the scale and orientation of the proposal distribution during the evolution to the posterior distribution. PSO is a stochastic optimization technique based on the movement and intelligence of swarms. In it a number of agents (i.e., particles) constitute a swarm that move around in the search space looking for the best solution. ACCO is a randomized search algorithm that is used in this study to function as an efficient sampler. We present the results showing, how well different algorithms estimate parametric uncertainty of hydrological models for a real field case of the Nzoia catchment in Kenya. We found that differences in sampling lead to quite large differences in the posterior distributions and hence different prediction intervals for the same model/problem considered. This effect has to be taken into account in the comparative studies of uncertainty analysis methods.

Chapter 6
Prediction of uncertainty by machine learning techniques

This chapter present an approach to predict the parametric uncertainty of a hydrological model. The most widely used methods to analyze uncertainty (which involve Monte Carlo (MC) simulation) involve multiple model runs and are time consuming. Machine learning techniques can be used to encapsulate the results of MC simulations and to build models that predict model uncertainty for future hydrological model runs. These predictive models are fast and can be easily used in operation for real-time predictions of parameter uncertainty. This method is referred to as Machine Learning in parameter Uncertainty estimation (MLUE) (Shrestha et al.2009; 2013)[4].

6.1 Introduction

As mentioned in the previous chapter, that Monte Carlo (MC) simulation could be unfeasible for computationally intensive models because of the associated requirements of time and resources. This method becomes impractical in real time applications when there is a time limitation to perform the uncertainty analysis, because of the large number of model runs required.

The number of simulations increases exponentially with the dimension of the parameter vector $O(np)$, where p is the dimension of the parameter vector, and n is the number of samples required for each parameter. Therefore, sampling from the selected distribution using standard MC simulation might be efficient.

The assessment of model uncertainty when it is used in operation is not widely discussed in the literature. The MC simulation provides only the averaged uncertainty estimates based on past data, but in real-time forecasting situations, there may be little time to perform the MC simulations for the new input data in order to assess the model uncertainty for a new situation. To overcome these problems, Shrestha et al. (2009, 2013) proposed to use machine learning techniques to emulate the MC simulation results obtained for the past data. They referred to this method as the MLUE (**M**achine **L**earning in parameter **U**ncertainty **E**stimation).

The idea of MLUE is to use the data from MC simulations to train a statistical or machine learning model to predict (with specially selected inputs) the quantiles and *pdf* of the model

---

[4] Shrestha, D. L., Kayastha, N., Solomatine, D. P. and, Price R. K. (2013) Encapsulation of parametric uncertainty statistics by various predictive machine learning models: MLUE method, *Journal of Hydroinformatics*, 16, 1, 95–113

Solomatine, D. P., Shrestha, D. L., Kayastha, N., Di Baldassarre, G. (2012). Application of methods predicting model uncertainty in flood forecasting, *Proc. Second European conference on FLOODRisk*, Rotterdam, The Netherlands.

error distribution. It requires only a single set of MC simulations in offline mode and allows one to predict the uncertainty bounds of the model prediction when the new input data are observed and fed into hydrological models (whereas the standard MC approach requires new multiple model runs for each new input).

The MLUE approach involves using machine learning techniques to encapsulate the information about the realisations of the hydrological model output generated by MC simulations. This approach emulates the complex model by using a simple model, which is an example of surrogate modelling, or meta-modelling - an approach that is widely used when running the complex model is computationally expensive. For example, O'Hagan (2006) used the Gaussian process emulator to emulate a complex simulation model. Li et al. (2006) proposed meta modelling whereby a sequential technique is used to construct and simultaneously update mutually dependent meta-models for multiresponse, high-fidelity deterministic simulations. Young and Ratto (2009) proposed a dynamic emulation model to emulate a complex high-order model by using a low-order data-based mechanistic model. The uniqueness of MLUE method is that it explicitly builds an emulator for the MC uncertainty results based on machine learning techniques.

This chapter presents the experiment of MLUE usage in parameter uncertainty estimation of a hydrological model. This method was employed and compared with different machine learning models (ANN, MT and LWR) to emulate MC simulation results . This methodology is tested with the GLUE uncertainty analysis method (see detailed description in Section 5.3.2) using HBV hydrological models for the Brue catchment in the U.K. and the Bagmati catchment in Nepal. The main results are presented in publication by Shrestha et al. (2013).

6.2 Machine learning techniques for building predictive uncertainty models

The machine learning techniques are briefly described in Chapter 2.5. Usually, predictive models built by machine learning are used for deterministic prediction. The main advantage of machine learning techniques is that they can build predictive models from data without knowledge of the internal system.

A detailed description of the MLUE method can be found in Shrestha et al. (2009, 2013) and is briefly outlined here. Instead of predicting a single value of the model error, which is done in most error correction procedures, it predicts the distribution of the output generated by MC-based simulations. Thus, the method predicts the uncertainty bounds of the hydrological model prediction without re-running the MC simulations. However, the MC-based uncertainty analysis methods require a fresh set of MC runs for each analysis. For instance, GLUE will typically require a fresh set of MC runs from the behavioural models to produce the prediction intervals for the model output for each time step with the new data input.

The flow chart of the MLUE methodology is presented in Figure 6-1. The basic idea here is to estimate the uncertainty of the model M (e.g., hydrological) output, assuming that uncertainty at a particular time step depends on the corresponding forcing input data and the model states (e.g., rainfall, antecedent rainfall and soil moisture) . In MC simulation, the vectors of parameters or inputs are sampled and for each of them, the hydrological model M

is run generates a time series of the model output $\hat{y}$. The results are presented in the matrix form $Y = \{\hat{y}_{t,s}\}$, where $t = 1,\ldots, N$, $s = 1,\ldots, S$, N is the number of time steps, S is the number of simulations. Note that each row of the matrix Y corresponds to the particular forcing vector x'_t, which is given by:

$$\{\hat{y}_{t,1},\ldots\ldots,\hat{y}_{s,1}\} = \{M(x'_t, \theta_1),\ldots\ldots, M(x'_t, \theta_s)\}$$

$$(6\text{-}1)$$

where θ is the parameter vector of the model M. Similarly each column of matrix Y, i.e. $\{\hat{y}_{1,s},\ldots,\hat{y}_{t,s}\}^T$ is one realisation of MC simulations corresponding to the parameter set θ_s. The machine learning model U is built to encapsulate MC results in the following form: $\hat{y}_t = U(x_t)$.

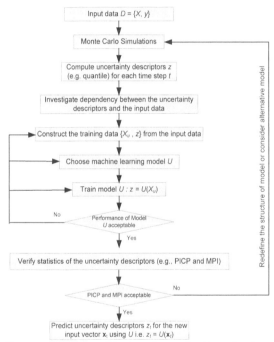

Figure 6-1. Schematic diagram of using machine learning method to estimate uncertainty generated by MC simulations (Shrestha et al. 2013).

In Figure 6-1, $z = \{z_1,\ldots, z_K\}$ is a set of desired statistical properties and x is the input vector of the model U, which is constructed from the forcing input variables x', model state s and possibly model output $\hat{y}$ (all possibly combined, transformed and/or lagged). A way to construct the input space x is described in Section 6.2.3. To characterise the uncertainty of the model M prediction the uncertainty descriptors should be considered, which are given in section below.

6.2.1 Characterization of uncertainty

Following MC simulations (see Chapter 5 Section 5.3.1), the quantiles of the model output for each time step t are estimated from the realizations $\hat{y}_{t,i}$. The uncertainty descriptors are characterized from model outputs, which are given below:

1. The prediction variance $\sigma_t^2(\hat{y}_t)$

$$\sigma_t^2(\hat{y}_t) = \frac{1}{s-1}\sum_{i=1}^{s}(\hat{y}_{t,i} - \bar{\hat{y}}_{t,i})^2 \tag{6-2}$$

where $\bar{\hat{y}}_{t,i}$ is the mean of the MC realizations at the time step t.

2. The prediction quantile $\hat{Q}_t(p)$ corresponding to the pth [0, 1] quantile calculated by Equation (5-4).

3. The transferred prediction quantile $\Delta\hat{Q}_t(p)$ corresponding to the pth [0, 1] quantile

$$\Delta\hat{Q}_t(p) = \hat{Q}_t(p) - \bar{y}_t \tag{6-3}$$

where $\bar{y}_t$ is the output of the calibrated (optimal) model. Note that the quantiles $\Delta\hat{Q}_t(p)$ obtained in this way are conditional on the model structure, inputs and the likelihood weight vector w_i. The essence of this transferred prediction quantile will be apparent in the next section.

4. The prediction intervals $[PI_t^L(\alpha) \quad PI_t^U(\alpha)]$ derived from the transferred prediction quantile for given confidence level of 1-α (0<α<1)

$$PI_t^L(\alpha) = \Delta\hat{Q}_t(\alpha/2), \quad PI_t^U(\alpha) = \Delta\hat{Q}_t((1-\alpha)/2) \tag{6-4}$$

where $PI_t^L(\alpha)$ and $PI_t^U(\alpha)$ are the distance between the model output to the lower and upper prediction limits respectively, and refer to the lower and upper prediction intervals corresponding to the 1-α confidence level (although, formally, these are not intervals but rather distances).

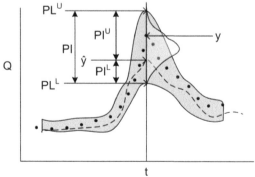

Figure 6-2. Prediction intervals as uncertainty descriptors (grey is uncertainty bound of streamflows, yellow is *pdf* at time t, black dot is observed streamflows, and black dash is simulated streamflows by hydrological model).

6.2.2 Techniques for building predictive uncertainty models

Once the desired uncertainty descriptors are computed from the realizations of MC simulations. The machine learning models are used to map the input data to the uncertainty descriptors. The machine learning model U to learn the functional relationship between the input data $\mathbf{x}$ and the uncertainty descriptor z takes the form:

$$z_t = U(\mathbf{x}_t) + \xi_t \tag{6-5}$$

where ξ_t is the residual (error) between the target uncertainty descriptor z_t and the predicted uncertainty descriptor by the machine learning model. The input data vector $\mathbf{x}$ used to train the machine learning models is typically different from the input to the process based model M and is discussed in Section 6.2.3. The input data $\mathbf{x}$ is constructed from the input variables of the process model, state variables, lagged variables of input and output, and other relevant variables that could help to increase the accuracy of the prediction. The residual ξ measures the accuracy and predictability of the machine learning model U. After being trained of model U, it encapsulates the underlying dynamics of the uncertainty descriptors of the MC simulations and maps the input to those descriptors. The model U can be used ANN, MT and LWR.depending on the complexity of the problem to be solved using available of data. Once the model U *is* trained on the calibration data, it can be employed to estimate the uncertainty descriptors such as prediction intervals for the new input data vector that were not used in any of the model building processes (see Figure 6-1).

If the uncertainty descriptor is the transferred prediction quantile (Equation (6-6)), then the model U will take the form:

$$\Delta \hat{Q}_t(p) = U(\mathbf{x}_t) + \xi_t \tag{6-6}$$

If the uncertainty descriptor is the prediction interval derived from the transferred prediction quantile (Equation (6-4)), then the model U will take the form:

$$PI_t^l(\alpha) = U_L(\mathbf{x}_t) + \xi_L$$
$$PI_t^U(\alpha) - U_U(\mathbf{x}_t) + \xi_U \tag{6-7}$$

Since the transferred prediction quantile is derived from the existing value of the model simulations (Equation (6-3)), then the predictive quantile of the model output is accordingly estimated by:

$$\hat{Q}_t(p) = U(\mathbf{x}_t) + \bar{y}_t \tag{6-8}$$

Similarly, upper and lower prediction limits of the model output are given by

$$PL_t^L(\alpha) = U_L(\mathbf{x}_t) + \bar{y}_t$$
$$PL_t^U(\alpha) = U_U(\mathbf{x}_t) + \bar{y}_t$$

(6-9)

where U_L and U_U are the machine learning models for the lower and upper prediction intervals, respectively. It is worthwhile to mention that for the uncertainty descriptors of Equation (6-3) and Equation (6-4), it is assumed that there is an optimal model.

6.2.3 Selection of input variables for the predictive uncertainty model

The selection of appropriate model inputs is extremely important as they contain important information about the complex (linear or non-linear) relationship with the model outputs. Therefore the success of the MLUE method depends on the appropriate selection of the input variables to use in the machine learning model U. The required input variables can vary depending on the type of the process model and the inputs used in the process model, among others. In most cases, a combination of the domain knowledge and the analytical analysis of the causal relationship may be used to select relevant variables to use as the input to the machine learning model. The input variables to the machine learning model that can be considered are termed *plausible data* and include:

1. Input variables to the process or primary model;

2. State variables;

3. Observed outputs of the process model;

4. Time derivatives of the input data and state variables of the process model;

5. Lagged variables of input, state and observed output of the process model; and

6. Other data from the physical system that may be relevant to the *pdf* of the model errors.

In most practical cases, the input data set $\mathbf{x}$ can be constructed from the plausible data set according to the methods discussed in section 2.5.6. Since the natures of the models M and U are very different, an analytical technique such as linear correlation or average mutual information between the quantiles of the model error and the plausible data is required to select the relevant input variables. As noted in the above list of the plausible data, the input variable might also consist of the lagged variables of input, state and observed output of the process model. Based on the domain knowledge and the analytical analysis of the causal relationship, several structures of input data can be tested to select the optimal input data structure.

For example, if the model M is a conceptual hydrological model, it would typically use rainfall (R_t) and evapotranspiration (E_t) as input variables to simulate the output variable runoff (Q_t). However, the uncertainty model U, whose aim is to uncertainty of the simulated runoff, may be trained with the possible combination of rainfall and evapotranspiration (or effective rainfall), their past (lagged) values, the lagged values of runoff, and, possibly, their combinations.

Let us recall Equation (6-5) of the uncertainty model U and extend it:

$$z_t = U(R_t^{'}, EP_t^{'}, Q_{t-1}^{'}, S_t^{'}, ...) + \xi_t \tag{6-10}$$

where,

$R_t^{'} = R_t, R_{t-1}, ..., R_{t-\pi max}$ is the lagged inputs of the rainfall

$EP_t^{'} = EP_t, EP_{t-1}, ..., EP_{t-\pi max}$ is the lagged inputs of the potential evapotranspiration

$Q_{t-1}^{'} = Q_{t-1}, Q_{t-2}, ..., Q_{t-\pi max}$ is the lagged inputs of the runoff

$S_t^{'} = S_t, S_{t-1}, ..., S_{t-\pi max}$ is the lagged inputs of the state variable (e.g., soil moisture etc)

The difficulty here is to select an appropriate lags πmax for each input variables beyond, which the values of the input time series have no significant effect on the output time series (in our case uncertainty descriptors). A subset of inputs for the model U is selected based on methods (CoC and AMI) discussed in Section 2.5.6. It is noteworthy to mention that inputs to the model U should not include those variables that are not available or cannot be measured at the time of prediction. Thus, in the above formulation, Q_t is not included. However, lagged of Q_t can be used as one of the inputs as shown above.

6.2.4   Verification of the predictive uncertainty models

The uncertainty model U can be validated its predictive capability; and measuring the statistics of the uncertainty. The former approach measures the accuracy of uncertainty models in approximating the uncertainty descriptors of the realizations of the MC simulations. The latter approach measures the goodness of the uncertainty models as uncertainty estimators. The coefficient of correlation (CoC) and the root mean squared error ($RMSE$) are used to measure the predictive capability of the uncertainty model and are given as:

$$CoC = \frac{\sum_{t=1}^{n}(z_t - \bar{z})(V(\mathbf{x}_t) - \bar{V}(\mathbf{x}_t))}{\sqrt{\sum_{t=1}^{n}(z_t - \bar{z})^2} \sqrt{\sum_{t=1}^{n}(V(\mathbf{x}_t) - \bar{V}(\mathbf{x}_t))^2}} \tag{6-11}$$

$$RMSE = \sqrt{\frac{1}{n}\sum_{t=1}^{n}(z_t - V(\mathbf{x}_t))^2} \tag{6-12}$$

where $\bar{z}$ and $\bar{V}(\mathbf{x}_t)$ are the mean of the uncertainty descriptors and the mean of the uncertainty descriptors predicted by the uncertainty model U, respectively. In addition to these numerical measures, the graphical plots such as the scatter and time plot of the uncertainty descriptors obtained from the MC simulation and their predicted values are used to judge the performance of the uncertainty model U.

The goodness of the uncertainty models is evaluated based on uncertainty measures prediction interval coverage probability ($PICP$) and mean prediction interval (MPI). Their equations are given in Section 5.4.2 (Equations 5-6 and 5-7). The $PICP$ is the frequency of the observed outputs falling within the computed prediction intervals corresponding to the prescribed confidence level of 1-α (say 90%). Thus, *PICP* measures the efficiency to bracket the observed outputs within the uncertainty bounds against the specified α value. Theoretically, the value of *PICP* should be close to the prescribed degree of confidence 1-α. However, if the value of *PICP* obtained by MC simulation is not close to 1-α, then we cannot expect that the *PICP* value obtained by uncertainty model *U* will be close to α. If the model *U* is sufficient good then its *PICP* value should be close to that of the MC simulations. *MPI* is the average width of the prediction intervals and provides an indication of the magnitude of the uncertainty. The larger the uncertainty, the larger the value of the *MPI* will be. In the ideal case that there is no uncertainty, then the value of the *MPI* will be zero. However, the *MPI* value alone does not provide much information; it will be used together with *PICP* to compare the performance of the uncertainty models. The best model will be that one that yields the *PICP* close to 1-α with the lowest *MPI*. It is obvious that the *PICP* will be increased with the increase of the *MPI*.

In addition to these uncertainty statistics, the plot of uncertainty bounds and the observed model output are investigated to judge the performance of the uncertainty model. Visual inspection of these plots can provide significant information about how effective the uncertainty model is in enclosing the observed outputs along the different input regimes (e.g., low, medium or high flows in hydrology).

6.3 Experimental setup

6.3.1 Uncertainty analysis for case studies Bagmati and Brue

In present chapter, the MLUE approach was used to test two catchments Brue and Bagmati. The descriptions of these catchments are provided in Section 1.4.3 and 1.4.2. The size of the Bagmati catchment and the size of the associated data set are larger than the Brue,. HBV hydrological model is calibrated by using adaptive cluster covering (ACCO) (Solomatine 1999), a global optimization method implemented in using GLOBE software. MC simulation results considering the only parameter uncertainty are not free from these sources of error. Although we try to reduce such errors, uncertainty results only considering parameter uncertainty. We explicitly consider only parameter uncertainly in this study.

The convergence of MC simulations is assessed to determine the number of samples required to obtain the reliable results (e.g. refer to Chapter 5 Section 5.5). The parameters of the HBV model are sampled using non-informative uniform sampling without prior knowledge of individual parameter distributions other than a feasible range of values (see Table 2-2 and Table 2-3). The likelihood measure is calculated based on the sum of the squared error in Equation (5-12), which corresponds to the *NSE*. The threshold value of *NSE* equal to 0 is selected to classify simulation as either behavioural or non-behavioural. The number of behavioural models is set to 25,000, which is based on the convergence analysis of MC simulations. Various uncertainty descriptors such as variance, quantiles, prediction intervals and estimates of the probability distribution functions are computed from these 25,000 MC realisations. Note that these descriptors are computed using the likelihood measure (Equation 5-12) as weights w_s in Equation (5-4). The model parameters ranges used for MC sampling are given in Table 2-3. For the Bagmati catchment, first 122,132 MC samples are generated by setting threshold value of 0.7 to obtain 25,000 behavioural samples.

However, to make consistent with the Brue catchment experiment, model simulations with negative *NSE* are removed from the further analysis, leaving 116,153 samples out of 122,132.

6.3.2 Machine learning models (ANN, MT and LWR)

A multilayer perceptron neural network with one hidden layer is used; the Levenberg-Marquardt algorithm is employed for its training. The hyperbolic tangent function is used for the hidden layer, and the linear transfer function is employed for the output layer. The maximum number of epochs is fixed to 1000. The trial-and-error method is adopted to find the optimal number of neurons in the hidden layer; we explored the number of neurons ranging from 1 to 10. It was found that 7 and 8 neurons for lower and upper PI, respectively yield the lowest CV error for the Brue catchment, For the Bagmati catchment, the number of hidden neurons 5 and 7 were used to yield the lowest CV error.

Experiments with MT are carried out with various values of the pruning factor that controls the complexity of the generated model (i.e., number of the linear models) and hence the generalizing ability of the model. We report the results of the MT which has a moderate level of complexity. The CV data set has not been used in the MT, rather it uses the whole calibration data set to build the model.

In the LWR model, two important parameters, number of neighbours and the weight functions are used. Several experiments were conducted with different combination of these parameter values. The best results were obtained using 5 neighbours and the linear weight function for the Brue catchment, and 11 neighbours with the Tricube weight function for the Bagmati catchment.

The selection of input variables for the machine learning model U is based on the methods outlined in Section 2.5.6. They are constructed from the forcing input variables (e.g. rainfall, evapotranspiration) used in the process models, and the observed discharge. The selected input variables are RE_{t-9a}, Y_{t-1}, ΔY_{t-1} for the Brue catchment and RE_{t-0}, RE_{t-1}, Y_{t-1}, Y_{t-2} for the Bagmati catchment. In the terminology for these variables, $RE_{t-\tau}$ is effective rainfall at time $t -\tau$, $Y_{t-\tau}$ is discharge at time $t -\tau$, τ is lag time, and RE_{t-9a} is the average of RE_{t-5}, RE_{t-6}, RE_{t-7}, RE_{t-8}, RE_{t-9} and ΔY_{t-1} is $Y_{t-1} - Y_{t-2}$. Due to the resolution of data is daily for the Bagmati catchment (as opposed to hourly data for the Brue), we do not consider the derivative (stepwise difference) of the flow as an input to the model for it.

The same data sets used for calibration and verification of the HBV model are used for the training and verification of model U, respectively. However, for proper training of the machine learning models, the calibration data set is segmented into two subsets: 15% of the data sets are used for CV and the remaining 85% are utilised for training. The CV data set was used to identify the best structure of machine learning models.

6.3.3 Modelling the probability distribution function

Shrestha et al. (2009) estimated the 90% prediction intervals (PIs) by building only two models predicting the 5% and 95% quantiles. However, Shrestha et al.(2013) extended that work to predict several quantiles of the model outputs to estimate the distribution functions (*CDF*) of the model outputs generated by the MC simulations. They shown that MLUE methodology estimate the two quantiles can be extended to approximate the full distribution of the model outputs. The procedures to estimate the *CDF* of the model outputs consist of: (i) deriving the *CDF* of the realisations of the MC simulations in the calibration data; (ii) selecting several quantiles of the *CDF* in such a way that these quantile can approximate the *CDF*; (iii) computing corresponding prediction quantiles using Equation (6-5); (iv) constructing and training separate machine learning models for each prediction quantiles; (v) using these models to predict the quantiles for the new input data vector; and (vi) constructing a *CDF* from these discrete quantiles by interpolation. This *CDF* will be an approximation to the *CDF* of the MC simulations.

In this study, we first estimated the 90% PIs by building only two models predicting the 5% and 95% quantiles, and later we extended the 19 quantiles from 5% to 95% with uniform interval of 5%. Next, an individual machine learning model was constructed for each quantile using the same structure of the input data and the model that was used for modelling two quantiles. In principle, the optimal set of inputs data and the model structure could differ for each quantile but we leave this investigation for future studies.

6.4 Results and discussion

The HBV model is calibrated by maximizing *NSE*. These maximized *NSE* values of 0.96 and 0.83 are obtained for the calibration period in the Brue and Bagmati catchments, respectively. The model is validated by simulating the flows for the independent verification data set, and *NSE* is 0.83 and 0.87 in the Brue and Bagmati catchment, respectively.

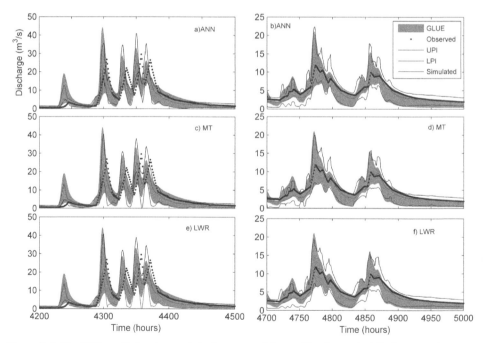

Figure 6-3. The 90% prediction intervals in fragment of the verification period for Brue catchment (LPI and UPI denote the prediction uncertainty estimated by ANN, MT, and LWR)

The HBV model is quite accurate for the Brue catchments but its error (uncertainty) is quite high during the peak flows for the Bagmati catchment, the standard deviation of the observed discharge in the validation period is 54% higher than that in the calibration period, which apparently increases performance during the validation period.

Figure 6-3 shows a comparison of the 90% prediction bounds estimated by the GLUE and three machine learning models (ANN, MT and LWR) in the verification period for the Brue catchment. One can see a noticeable difference among them for predicting the lower and upper bounds of PI. For example, in the second peak of Figure 6-3 4a, the upper bound of PI is underestimated by the ANN compared with the MT and LWR. However, the lower bound is well approximated by the ANN as compared with the other models. Furthermore, in Figure 6-3b, the ANN is overestimating two peaks, while the MT and LWR models underestimate them (Figure 6-3d and f). From the Figure 6-3, it can be seen that the results of the three models are comparable. They reproduce the MC simulations uncertainty bounds reasonably well except for some peaks, in spite of the low correlation of the input variables with the PIs. The predicted uncertainty bounds follow the general trend of the MC uncertainty bounds although some errors can be noticed and the models fails to capture the observed flow during one of the peak events (Figure 6-3a, c, and e).

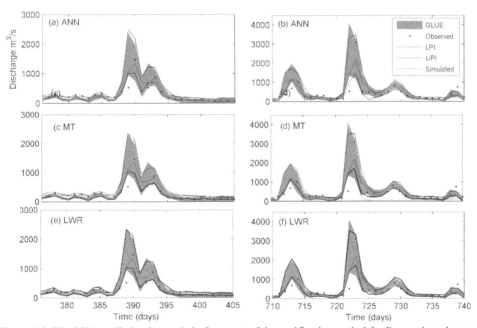

Figure 6-4. The 90% prediction intervals in fragment of the verification period for Bagmati catchment (LPI and UPI denote the prediction uncertainty estimated by ANN, MT, and LWR)

For the Bagmati catchment, it is found that only 49.79% of the observed discharge data are inside the 90% prediction bounds computed by the GLUE method in the calibration period and 61.48% in the verification period. Therefore, we follow the modified GLUE method (denoted by mGLUE) (Xiong and O'Connor, 2008) to improve the capacity of the prediction bounds to capture the observed runoff data. The mGLUE method uses the bias-corrected MC simulations to estimate the uncertainty bounds. Compared with the original GLUE method (Beven and Binley, 1992), the mGLUE method includes two additional procedural steps.

Firstly, for each behavioural parameter set, a simulation bias curve is constructed based on the simulation series that are obtained using the calibration data. Thus, for a number of behavioural parameter sets, there will be different simulation bias curves. Secondly, at each time step, with the new data input, all of the different prediction values for the same observation are corrected by dividing by a common median bias value, before the derivation of the prediction limits. In further reading, the mGLUE term is referred as "GLUE" for the Bagmati catchment.

Figure 6-4 presents the 90% prediction bounds estimated by the GLUE and the three machine learning models in the verification period. Using the GLUE method, the percentage of the observations falling inside the bounds is increased to 65.26 % and 67.52% in the calibration and verification periods, respectively. The machine learning models are able to approximate the GLUE simulation results reasonably well. The results of the three machine learning models are comparable; however, one can see a noticeable difference between them when predicting the peaks. The highest peak is overestimated by the ANN model, while the other two peaks are underestimated.

Figure 6-5 and Table 6-1 present a summary of statistics of the uncertainty estimation in the verification period. The ANN model is very close to the MC simulations results. The MT and LWR are better than the ANN with respect to MPI (note that lower MPI indicates better performance), however $PICP$ shows that the prediction limits estimated by them enclose relatively lower percentage of the observed values compared with those of the ANN.

Table 6-1. Performances of the models measured by the coefficient of correlation (CoC), root mean squared error ($RMSE$), the prediction interval coverage probability ($PICP$) and the mean prediction interval (MPI).

Catch-ments	Models	Lower prediction interval				Upper prediction interval				PICP		MPI	
		Calibration		Verification		Calibration		Verification					
		CoC	RMSE	CoC	RMSE	CoC	RMSE	CoC	RMSE	Cal.	Ver.	Cal.	Ver.
Brue	ANN	0.91	0.70	0.86	0.56	0.80	1.61	0.80	1.59	90.03	77.00	2.73	2.09
	MT	0.91	0.77	0.84	0.61	0.81	1.62	0.79	1.63	84.24	68.72	2.54	1.95
	LWR	0.91	0.78	0.82	0.64	0.78	1.73	0.80	1.60	89.33	75.43	2.54	1.93
Bagmati	ANN	0.87	38.51	0.81	51.46	0.97	37.69	0.94	61.59	60.26	66.24	122.12	124.03
	MT	0.83	44.17	0.81	50.25	0.96	48.27	0.95	52.14	58.92	59.05	117.82	120.59
	LWR	0.91	32.43	0.86	44.56	0.97	40.11	0.96	50.37	63.04	59.16	118.91	121.73

We have compared the performance of the three machine learning models by analyzing the accuracy of the prediction as well as there are other factors to be considered. These include computational efficiency, ease of use, number of training parameters required, flexibility and transparency. These considerations are shown in Table 6-2. The time (hh:mm:ss) is based on prediction of two quantiles (5% and 90%). Data analysis includes the analysis of dependency between prediction intervals and the input data, and preparation time in the calibration period except GLUE; The computer used for this study is an Intel (R)

120

Pentium(R) Dual CPU E2160 @ 180 GHz 2GB RAM. We use linguistic variables to describe these factors. It can be observed that none of the models is superior with respect to all factors; however, one may favour the ANN if the ranking is done by assigning equal weight to all factors.

Table 6-2. Computational time for GLUE and MLUE

Catchments		Brue		Bagmati	
Periods		Calibration	Verification	Calibration	Verification
Number of data use		8760	8217	2000	922
GLUE	Monte Carlo simulations	15:49:00	11:12:00	07:35:00	06:36:00
	Estimation of quantiles	00:45:00	00:33:00	00:10:00	00:05:00
		16:34:00	**11:45:00**	**07:45:00**	**06:41:00**
ANN	Data analysis	00:05:00		00:02:00	
	Preparation of training data	00:02:00	00:02:00	00:01:00	00:01:00
	Train the model	02:00:00		01:00:00	
	Testing the model		00:02:00		00:00:30
		02:07:00	**00:04:00**	**01:03:00**	**00:01:30**
MT	Data analysis	00:05:00		00:02:00	
	Preparation of training data	00:02:00	00:02:00	00:01:00	00:01:00
	Train the model	01:00:00		00:30:00	
	Testing the model		00:01:00		00:00:05
		01:07:00	**00:03:00**	**00:33:00**	**00:01:05**
LWR	Data analysis	00:05:00		00:02:00	
	Preparation of training data	00:02:00	00:02:00	00:01:00	00:01:00
	Train the model	04:00:00		02:00:00	
	Testing the model		00:07:00		00:02:00
		04:07:00	**00:09:00**	**02:03:00**	**00:03:00**

6.4.1 Comparison among ANN, MT and LWR

Figure 6-6 and Figure 6-7 show a comparison of the *CDF*s for the peak events estimated by the three machine learning methods for the Brue and Bagmati catchment, respectively. One can see that the *CDF*s estimated by the ANN, MT and LWR are comparable and are very close to the *CDF*s given by the MC simulations. It is observed that the *CDF*s estimated by the ANN, MT and LWR models deviate a slightly near the peak event of 9 January 1996 in the Brue catchment (see Figure 6-6b). The *CDF*s estimated by the ANN, MT and LWR deviate a bit more at the higher percentile values for the peak event of 13 August 1995 in the Bagmati catchment (see Figure 6-7b).

From the visual inspection the plot, one can see that the *CDF*s are reasonably approximated by the machine learning methods. However, it may require a rigorous statistical test to conclude if the estimated *CDF*s are not significantly different from those given by the

MC simulations. In this study, since we have limited data (only 19 points) the results of the significance test (e.g. Kolmogorov-Smirnov) may not be reliable.

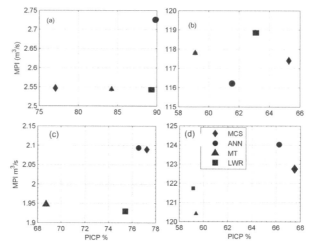

Figure 6-5. The comparison of *PICP* and *MPI* estimated with GLUE, ANN, MT, and LWR) (a) and (c) calibration and verification for the Brue catchment, (b) and (d) for the Bagmati catchment

Table 6-3. Linguistic performance criteria of machine learning models

Models	Model parameters (optimized)	Accuracy		Efficiency	Transparency	Rank
		CoC	PICP and MPI			
ANN	Numbers of hidden nodes	High	High	Medium	Low	1
MT	Pruning factor	Medium	Low	High	Medium	2
LWR	Kernel weighted density function	Low	Medium	Low	High	3

In this study, the uncertainty of the model output is assessed when the hydrological process model is used in simulation mode. However, this method can also be used in forecasting mode, provided that the process model is also run in forecasting mode. Note that we have not used the current observed discharge Q_t as an input to the machine learning models because this variable is not available during the model application (indeed, the value of this variable is calculated by the HBV model, and the machine learning model assesses the uncertainty of this output).

The results shown that the performance of the machine learning models to predict lower quantiles (5%, 10%, etc.) is relatively higher compared with those of the models for the upper quantiles (95%, 90%, etc.). This can be explained by the fact that the upper quantiles correspond to higher values of flow (where the HBV model is obviously less accurate) and higher variability, which makes prediction a difficult task. It is possible to develop a specific model only to simulate the peak observed data and their uncertainty, as well as for the mean flows. In general, such a model performs better than the global model. In this study, we have used MT and LWR models for uncertainty estimation that implicitly builds the local models internally. It would be interesting to build the local models explicitly for high-flow events, but, this is not always possible because of training data requirements for such rare and extreme events.

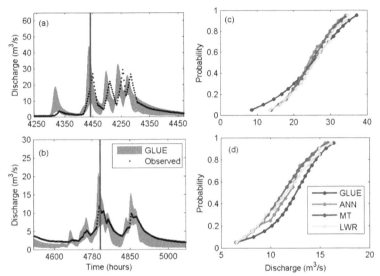

Figure 6-6. A comparison of cumulative distribution function (*CDF*) estimated with GLUE and ANN, MT, and LWR for the Brue catchment in a part of the verification period. (a) peak event of 20 December 1995,(b) peak event of 09 January 1996, (c) *CDF* at (a) and (d) *CDF* at (b)

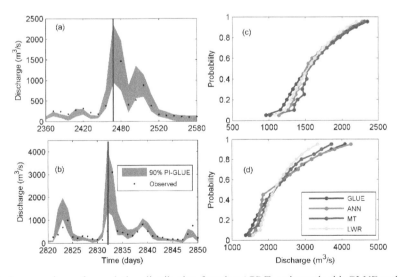

Figure 6-7. A comparison of cumulative distribution function (*CDF*) estimated with GLUE and ANN, MT, and LWR for the Bagmati catchment in a part of the verification period. (a) peak event of 14 September 1994, (b) peak event of 13 August 1995 (c) *CDF* at (a) and (d) *CDF* at (b)

When comparing the percentage of the observed discharge data falling within the uncertainty bounds (i.e. *PICP*) produced by the GLUE method, it can be seen that this percentage is much lower than the specified confidence level to generate these bounds. Low *PICP* value is consistent with the results reported in the literature (e.g., Montanari, 2005; Xiong and O'Connor 2008).

To approximate *CDF*, an individual machine learning model is constructed for each quantile with the same structure of the input data and the model configuration. Thus we have not undertaken the full-fledged optimization of the model and the input data structure of the machine learning models and there is a possibility to improve the results. Furthermore, one can notice that the *CDF*s estimated are not necessarily monotonically increasing (see, e.g. 30% quantile of the MT model for the second case study). This is not surprising given that individual models are built for each quantile independently. This deficiency can be addressed by a correcting scheme (to be developed) that would ensure monotonicity of the overall *CDF*.

6.5 Summary

This chapter presents the machine learning techniques to predict parameter uncertainty in hydrological modelling. The MLUE method is used to encapsulate the computationally expensive MC simulations of a process model by an efficient machine learning model. This model is first trained on the data generated by the MC simulations to encapsulate the relationship between the hydro-meteorological variables and the uncertainty descriptors – characteristics of the model output probability distribution, e.g. quantiles. The trained model can subsequently be used to estimate the latter for the new input data.

In MLUE, the three machines learning techniques, namely ANN, MT and LWR are used to predict several uncertainty descriptors of the hydrological model outputs. Two case studies demonstrated the application of the MLUE method. First, the method was tested to estimate the two quantiles (5% and 95%) forming the 90% PIs and later extended the 19 quantiles from 5% to 95% with a uniform interval of 5%, to approximate the *CDF* of the model outputs, and then an individual machine learning model is constructed for each quantile. Several performance indicators and visual inspection were used to evaluate MLUE. The results of the MLUE experiments showed that machine learning models are reasonably accurate to approximate the GLUE uncertainty bounds as well as estimate the *CDF* resulting from the GLUE. The MLUE method is computationally efficient and can be used in real-time applications when a large number of model runs are required.

Chapter 7
Committees of models predicting models' uncertainty

This chapter also presents machine learning methods in uncertainty prediction, but instead of using one model, we form a *committee* of predictive models. Artificial neural networks, model tree, and locally weighted regression are employed. Two committees of models are formed (i) six different model input structures for uncertainty prediction models; and (ii) seven different uncertainty prediction models from the results of different sampling based methods (MCS, GLUE, MCMC, SCEMUA, DREAM, PSO and ACCO). These models are combined to improve performances of their outputs. This approach is applied to estimate the uncertainty of streamflows simulation from a conceptual hydrological model in the Bagmati catchment in Nepal, and the Nzoia catchment in Kenya.[5]

7.1   Introduction

The motivation behind multi-model averaging is to extract as much information as possible from the existing competing models to produce better outputs. The analysis of the results from groups of competing models is much more complex than the analysis of any single model. Each model has its own predictive capabilities and limitations. Hence, it is difficult to compare effectively between models. However, the combination of competing models allows the strengths of each individual model to merge in an optimal way, so that the best prediction can be obtained.

Different hydrological models have various strengths in capturing different aspects of the hydrological processes, and different objective functions have advantages in simulating a veriety of flow ranges. Such model outputs can be combined into a single new model by employing the committee approach. The results of these models and their predictions are described in Chapters 3 and 4. Combining models requires weights, which average the model outputs, taking advantage of each individual model's strengths.

Uncertainty analysis is an essential component for any hydrological modelling effort. The sampling-based method is largely used to characterize and for quantifying the uncertainty of hydrological models. Machine learning techniques are used to encapsulate the results of Monte Carlo (MC) simulations by building a predictive uncertainty model (Shrestha et al., 2009). The machine learning-based uncertainty prediction approach is very useful for the estimation of hydrological models' uncertainty, in particular the hydro-metrological situation

---

[5] Kayastha, N., Solomatine, D. P, Shrestha, D. L., (2014) Prediction of hydrological models' uncertainty by a committee of machine learning-models, *11th International conference on Hydroinformatics*, New York USA

in real-time applications. In this approach, the hydrological model realizations from MC simulations are used to build different machine learning uncertainty models to predict uncertainty (quantiles of *pdf*) of the a deterministic output from a hydrological model. Uncertainty models are trained using antecedent precipitation and streamflows as inputs. The trained models are then employed to predict the model output uncertainty, which is specific for the new input data. This approach can be used for results of any sampling scheme to build machine learning models that are able to predict the uncertainty of hydrological model outputs. The trained model, called a predictive uncertainty model (V), maps the input data to the prediction interval of the model output, generated by sampling schemes. Details of this methodology are described in Chapter 6.

This chapter presents the results of hydrological model uncertainties predicted by several of machine learning models. Three machine learning models, namely artificial neural networks (ANNs), model tree (MT), and locally weighted regression (LWR) with (i) six different model input structures are tested to predict the uncertainty of streamflow simulations from a conceptual hydrological model an HBV for Bagmati catchment in Nepal; and (ii) seven different sampling-based uncertainty estimation methods applied to Nzoia catchment in Kenya. The problem here is that several input datasets were used to train model V resulting in a total of 18 models for the Bagmati and 21 models for the Nzoia. These are difficult to compare. In such situation, the multi-model averaging can be applied in order to combine these models. The main objective of the combining different predictive uncertainty models is to use efficiently the available information and to construct an averaged predictive uncertainty model with the proper balance between model flexibility and over-fitting. We propose to form a committee of all predictive uncertainty models using averaging schemes to generate the single (final) output. Two averaging schemes, namely simple averaging (SA) and Bayesian model averaging methods (BMA) are used in this study.

7.2 Bayesian Model Averaging

Bayesian Model Averaging (BMA) is a statistical technique used to combine multiple models for better prediction among various competing models. The main idea of BMA is that the ensemble outputs, which are generated by various models, are combined based on their performance. The better-performing models receive higher weights, so that the final combined model outputs can be much closer to on-the-ground observations. Using varying weights from one model to another makes more physical sense and decreases the uncertainty in the forecast (Ajami et al, 2007). We use BMA to combine the ensemble of predictive uncertainty models. The brief description of this method is described below.

The quantity Q to be predicted on the basis of input data $D=[X, Y]$. Where X denotes the input forcing data, and Y represents for the observed data. The ensemble of the k-member predictions is given as $f= [f1,f2,...,fk]$. The probabilistic prediction of BMA is given by:

$$p(Q|D) = \sum_{k=1}^{K} p(f_k | D) p_k(Q | f_k, D)$$
(7-1)

where, $p(f_k|D)$ is the posterior probability of the kth individual prediction f_k given input data D. This reflects how well model f_k fits Y, which was denoted as BMA weight w_k . Better performing predictions receive higher weights than the worse performing ones. All weights are positive and should sum to 1. $p_k(Q \mid f_k, D)$ is the conditional probability density function (PDF) of the predicting Q conditional on f_k and D. For convenience of computation, $p_k(Q \mid f_k, D)$ is assumed to be a normal *pdf* and is represented as $g(Q \mid f_k, \sigma_k^2) \sim N(f_k, \sigma_k^2)$, where σ_k^2 is the variance associated with model prediction f_k and observations Y.

The BMA mean prediction is a weighted average of the individual model's predictions, with their posterior probabilities being the weights. This quantity can be expressed as

$$E(Q \mid D) = \sum_{k=1}^{K} p(f_k \mid D) \cdot E\left[g(Q \mid f_k, \sigma_k^2)\right] = \sum_{k=1}^{K} w_k f_k \qquad (7\text{-}2)$$

The equation can be rewritten for combining multiple models of prediction intervals as follows

$$PI_B^{[L]} = \sum_{k=1}^{K} w_k V_k^L \qquad (7\text{-}3)$$

$$PI_B^{[U]} = \sum_{k=1}^{K} w_k V_k^U \qquad (7\text{-}4)$$

where, PI^L and PI^U are the distance between the model output to the lower and upper prediction limits, respectively, and refer to the lower and upper prediction intervals.

Expectation-Maximization (EM) algorithm is used to estimate BMA weight w_k and model prediction variance σ_k^2, based on the assumption that k-member predictions are normally distributed.

7.3   Building predictive uncertainty models for the Bagmati catchment

Machine learning models are separately employed for encapsulating the uncertainty estimated by the GLUE method, which is supposed to be fit for reliable uncertainty bound, thus forming two-member individual predictions (upper and lower prediction interval) for uncertainty predictions. Three machine learning models, namely ANN, MT, and LWR are used to build a predictive uncertainty model. The same data sets of the HBV model are used for the training and verification of the model V, respectively. However, for proper training of the machine learning models the calibration data set is segmented into two subsets: 15% of data sets for cross-validation (CV), and 85% for training (see., Figure 2-9). The CV data set is used to identify the optimal structure of the machine learning models.

The input variables for model V are constructed from rainfall and the observed discharge based on correlation and average mutual information (AMI). Experimental results show that evapotranspiration alone does not have a significant influence on the prediction intervals (PIs). Thus, it is not included as a input variable for model V but effective rainfall is used. Figure 7-1 shows the correlation coefficient and the AMI of RE_t and its lagged variables with the lower and upper prediction intervals (PIs). The optimal lag time (time at which the correlation coefficient and/or AMI is maximum) is also 0 and 1 hour. At this optimal lag time, the variable RE_t provides a maximum amount of information about the PIs. Additionally, correlation and AMI between the PIs and the observed discharge are analyzed. The results show that the immediate and the recent discharges (with the lags of 0, 1, 2 h) have very high correlation with the PIs. Consequently, it was also decided to also use the past values of the observed discharge as additional input to the model V.

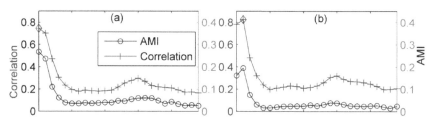

Figure 7-1. Simulated linear correlation between rainfall and (a) lower prediction interval; and (b) upper prediction interval.

Based on the above analysis, several structures of the input data for the machine learning models are considered. We use various combinations of three effective rainfall values denoted by RE_{t-0}, RE_{t-1} and RE_{t-2}, and past values of the observed discharge (see Table.7-1). The derivative of the flow indicates whether the flow situation is either normal or base flow (zero or small derivative), or can be characterized as the rising limb of the flood event (high positive derivative), or the recession limb (high negative derivative). Therefore, in addition to the flow variable, Q_{t-1}, the rate of the flow change at time $t-1$ is also considered.

Table.7-1. Input data structures of machine learning models to reproduce GLUE uncertainty results of the HBV model

Models	Input combination for lower and upper prediction interval
V01	RE_{t-0}, Q_{t-1}
V02	$RE_{t-0}, Q_{t-1}, \Delta Q_{t-2}$
V03	$RE_{t-0}, RE_{t-1}, Q_{t-1}, Q_{t-2}$
V04	$RE_{t-0}, RE_{t-1}, Q_{t-1}, \Delta Q_{t-1}$
V05	$RE_{t-0}, RE_{t-1}, RE_{t-2}, Q_{t-1}, Q_{t-2}$
V06	$RE_{t-0}, RE_{t-1}, RE_{t-2}, Q_{t-1}, \Delta Q_{t-1}$

Note: ΔQ_{t-1} is $Q_{t-1} - Q_{t-2}$ (characterizes the derivative of previous discharge)

In Table.7-1, six possible combinations of input structure considered for the machine learning models are presented. Note that these models are trained identically for lower and upper PI. However, it is possible to use a single model in some machine learning models (e.g., ANN) to produce two outputs given that it has the same input structures. The structure of the two machine learning models to estimate the lower and upper PI, for instance, in V03 configuration, takes the following forms:

$$PI_t^L = V_L(RE_{t-0}, RE_{t-1}, Q_{t-1}, Q_{t-2})$$
$$PI_t^U = V_U(RE_{t-0}, RE_{t-1}, Q_{t-1}, Q_{t-2})$$ (7-5)

where PI_t^L and PI_t^U are the lower and upper PI for the tth time step. Note that we have not used Q_t as an input to the above machine learning model because during the model application, this variable is not available (Indeed, the prediction of this variable is done by the HBV model, and the machine learning model assesses the uncertainty of the prediction). Furthermore, we would like to stress that in this study the uncertainty of the model output is assessed when the model is used in simulation mode. However, this method can also be used in forecasting mode, provided that the process model is also run in forecast mode.

7.3.1 Several sets of variables

We built six ANN uncertainty models (V) based on different input structures (Table.7-1) using a multilayer perceptron network with one hidden layer. Optimization is performed using the Levenberg-Marquardt algorithm. The hyperbolic tangent function is used for the hidden layer with linear transfer function at the output layer. The maximum number of epochs is fixed to 1000. A trial-and-error method is adopted to determine the optimal number of neurons in the hidden layer, in which a number of neurons from 3 to 10 for both lower and upper PI, which yielded the lowest error on the CV data set. The performance of the ANN model V for different input structures is shown in Table 7-2. It is observed that ANN models V03, V04, V05 and V06 possess almost similar CoC in the verification data set in producing upper PI. However, V01 or V02 have the smaller value of CoC. In lower PI, V01 and V02 gives the lowest; V03, V04 and V06 yield moderate CoC values; and V05 attains the highest CoC value in the verification dataset. V05 yields the best result corresponding to a $PICP$ value of 78.77% in comparison with the required 90%. However, the average width of the PIs (i.e., MPI) is wider. This is one of the reasons to cover additional observed data inside the PIs. V01 gives the lowest MPI value with the lowest $PICP$ value. The MPI values of V03 and V04 are very similar to the MC simulation results.

The performances of various MT models V using different input structures are shown in Table 7-2. MT models use calibration and verification data sets. A number of pruning factors 6 to 8 and 2 to 8 were used to build MT models for lower and upper PI, respectively. For upper PI, the model V03 was build using 8 numbers of linear models to obtain the best CoC values in calibration. The models V01 and V02 were produced low CoC values for upper PI, but these values still higher than values yielded by lower PI. The analysis reveals that 59.05% to 66.24% of the observed data were enclosed within the estimated 90% PIs in the verification period. The MPI values from all models are comparable. MT yeilds low value of the $PICP$ with lower value of MPI. Note that the $PICP$ and MPI value of the MC simulation results were 67.52% and 122.74 m³/s in the verification period.

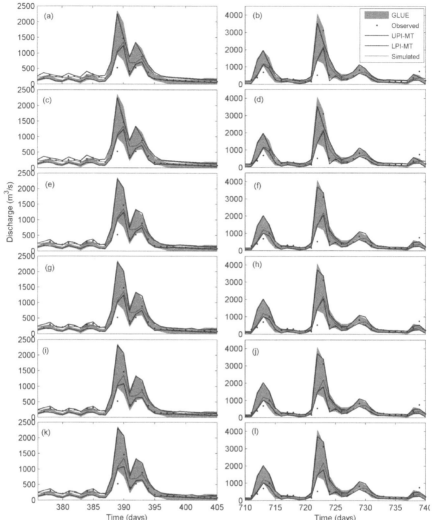

Figure 7-2. The fragments of 90% prediction intervals by GLUE and MT in verification period with (a and b) V01, (c and d) V02, (e and f) V03, (g and h) V04, (i and j) V05, (k and l) V06. (The black dot indicates observed discharges, the grey shaded area denotes the prediction interval by GLUE, the black line denotes the prediction interval by MT.

Figure 7-2 shows the comparison of 90% PIs estimated by MC simulations with 6 different input configurations for MT in the verification period. It was observed that upper PIs on some peak flows are underestimated by both models of MT (i.e., V01 and V02). One can see in the some peaks of upper PI in hydrographs (Figure 7-2 b, d, f, j, l) failed to cover by MT . The models V03, V04, V05, V06 (Figures 7-2 f, h, j and i), attempted to follow the upper prediction intervals estimated by GLUE, although some errors can be noticed. Noticeably, the models succeed to capture the observed flow during one of the peak events.

The two important parameters, namely the number of neighbours and the weight function, are considered to build LWR models. With different combinations of these values, several experiments were performed. The experiment was conducted with a number of neighbours, from 0 to 15, using weighted functions namely; linear, Epnechnikov and Tricube. We reported the best results obtained from 11 numbers of neighbour and Tricube weight function. The comparison of the performance of LWR by *CoC, RMSE, PICP* and *MPI*,with different input configurations is shown in Table 7-2. The models V03, V02, V03 and V04 have similar results, and these are slightly better than other models with respect to *CoC* values in the verification period. The *CoC* values for lower and upper PI are 0.86 and 0.96, respectively, for the V03 model. The experiment shows that the low value 0 to 1 of k-Nearest neighbour (KNN) unable to present good LWR models, only values greater than 2 started to improve the performances of these models. We also tested with different values of KNN, from 1 to 15, for the V03 model, which are shown in Figure 7-3. The value of KNN is 11 with Tricubic weighted kernel function is used for building different input structure of the LWR model.

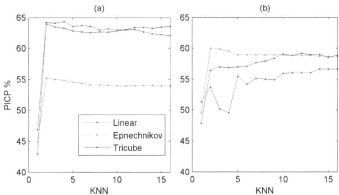

Figure 7-3. The performances of LWR models using different kernel functions and KNN values (a) calibration and (b) verification

7.3.2 Model averaging results and discussion

The BMA is applied for combining 18 individual predictive uncertainty models based on six different variants of input structures with three machine learning models (ANN, MT and LWR) for calibration and verification periods. These models were tested on data from the Bagmati catchment and it results are presented in Table 7-2. The outputs generated by various models (considering two quantities that arc lower and upper PI.) were combined using BMA. Each model (e.g., lower PI) received weights, which were calculated based on *CoC*, and then the final averaged model was compared by predictive measures of uncertainty, *PICP* and *MPI*. The result of the BMA model shows that the performance of *PICP* can be improved by individual models (MT and LWR models). ANN models showed performances that were between the best and the worst. The best of ANN models is V05, which has a value of a *CoC* value of 0.89 and 0.87 in calibration and verification, respectively. Its value of *PICP* is 74.43% in calibration and 78.77% in verification. The BMA yields 64.35 % and 69.74 % in the calibration and verification periods, respectively. However, it produced wider *MPI* among all models except ANN V05.

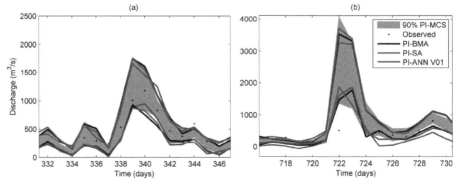

Figure 7-4. Hydrograph of 90% prediction bounds in the verification period; the black dot indicates observed discharges, and the dark grey shaded area denotes the prediction uncertainty that results from MCS. Black, blue and purple lines denote the prediction uncertainty estimated by BAM, SA and ANN-V01, respectively.

Table 7-2. Performances of the machine learning models and BMA

ML techni ques	Mod -els	CoC				RMSE							
		Lower PI		Upper PI		Lower PI		Upper PI		PICP		MPI	
		Cal	Ver	Cal	Ver	Cal	Ver	Cal	Ver	Cal	Ver	Cal	Ver
ANN	V01	0.83	0.71	0.89	0.86	44.57	60.25	74.85	88.05	55.26	56.84	117.81	118.73
	V02	0.85	0.71	0.90	0.86	41.56	60.79	76.89	92.09	70.62	75.52	141.84	142.80
	V03	0.85	0.81	0.96	0.94	41.98	51.46	47.80	61.59	60.26	66.24	122.12	124.03
	V04	0.86	0.81	0.95	0.94	40.10	49.96	50.08	60.81	60.52	68.91	124.56	125.79
	V05	0.89	0.87	0.96	0.95	36.29	43.34	61.45	67.53	74.43	78.77	160.96	160.48
	V06	0.88	0.82	0.95	0.93	37.78	49.54	54.52	66.28	68.35	73.32	135.56	136.94
MT	V01	0.77	0.72	0.88	0.90	49.77	59.14	77.03	76.92	57.32	64.04	117.68	118.95
	V02	0.78	0.73	0.88	0.90	48.91	58.68	76.67	76.81	58.66	66.24	117.70	119.14
	V03	0.87	0.77	0.95	0.95	38.96	54.93	49.61	53.14	58.87	59.40	117.80	120.42
	V04	0.87	0.76	0.95	0.95	39.16	55.66	49.22	53.27	59.07	60.09	117.74	119.67
	V05	0.83	0.81	0.96	0.95	44.17	50.25	48.27	52.14	58.92	59.05	117.82	120.59
	V06	0.83	0.80	0.96	0.95	44.29	51.18	47.99	52.21	59.12	59.51	117.76	119.89
LWR	V01	0.84	0.71	0.91	0.89	43.20	59.80	67.26	78.42	61.96	61.37	117.21	120.19
	V02	0.86	0.74	0.93	0.90	40.06	57.12	61.59	73.83	62.37	58.82	116.62	118.65
	V03	0.91	0.86	0.97	0.96	32.43	44.56	40.11	50.37	63.04	59.16	118.91	121.73
	V04	0.91	0.86	0.97	0.96	33.43	44.42	40.60	51.09	62.27	57.89	117.93	121.01
	V05	0.92	0.87	0.97	0.96	31.91	43.33	38.98	49.62	63.71	59.74	119.92	123.05
	V06	0.92	0.86	0.97	0.96	31.92	44.10	38.98	49.85	62.73	59.28	119.42	122.33
SA		**0.86**	**0.79**	**0.94**	**0.93**	**40.03**	**52.14**	**55.66**	**64.11**	**62.08**	**63.57**	**123.30**	**125.24**
BMA		**0.91**	**0.86**	**0.96**	**0.94**	**32.96**	**45.79**	**47.83**	**47.84**	**64.35**	**69.74**	**135.26**	**136.20**

7.4 Building predictive uncertainty models for the Nzoia catchment

The results Monte Carlo (MC) simulation-based estimation of model uncertainty depend on the sampling method used. In Chapter 5, we tested different sampling schemes (MCS, GLUE, MCMC, SCEMUA, DREAM, PSO and ACCO) for the uncertainty estimations of hydrological models. We used the results of these sampling schemes to build various machine learning models and these models were combined using simple averaging (SA) and Bayesian model averaging (BMA) methods, and tested their performances for the Nzoia catchment.

The uncertainty of model output generated by various sampling schemes is described by 90% prediction intervals.

Three machine learning models, namely ANN, MT, and LWR were separately employed to encapsulate the uncertainty results from each sampling-based method for uncertainty predictions. The experiment follows the results of previous experiments on sampling-based uncertainty. In Chapter 5, 10 years of daily data were considered as calibration data set (see Chapter 1, Section 3). Here, we presented a comparison of different predictive uncertainty models and their combinations in both calibration and verification periods. In addition, emphasis is placed to testing (verification). For proper training of the ANN machine learning model the calibration data set is segmented into two subsets: 15% of data sets for cross-validation (CV), and 85% for training to identify the optimal structure of the ANN models.

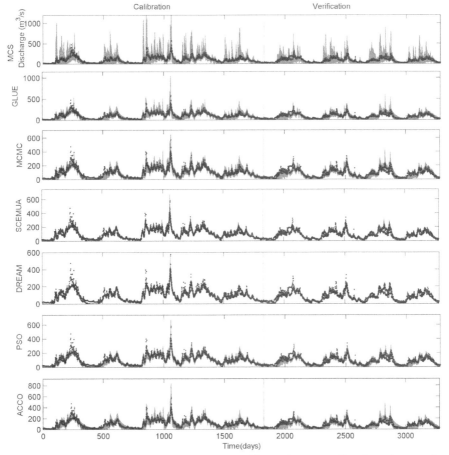

Figure 7-5. Prediction intervals generated by various sampling-based uncertainty estimation methods

Table.7-3.Performances of various sampling-based uncertainty estimation methods

Sampling Methods	PICP		MPI	
	Cal	Ver	Cal	Ver
MCS	82.73	85.22	180.48	163.45
GLUE	82.39	75.52	110.17	94.61
MCMC	72.08	62.24	76.81	66.95
SCEMUA	22.31	18.20	19.03	16.07
DREAM	9.40	7.07	7.13	6.17
PSO	70.67	58.46	60.37	54.97
ACCO	79.45	72.52	92.57	83.51
Average	78.14	78.11	70.44	70.21

Study area and selection of input variables

The input variables for model V were constructed from correlations and AMI, which also have been described in a previous Chapter, Section 6.2.3. Figure 7-6 shows the correlation coefficient and the AMI of RE_t and its lagged variables with the lower and upper PI. The optimal lag time (time at which the correlation coefficient and/or AMI is maximum) and the variable RE_t provide the maximum amount of information about the PIs. The immediate discharges with the lag of 1 and 2 have very high correlations with the PIs.

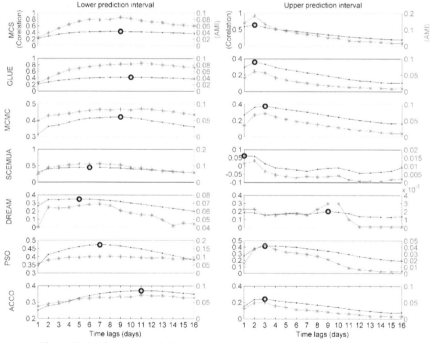

Figure 7-6. Linear correlation and average mutual information (AMI) between effective rainfall (a) lower prediction interval; and (b) upper prediction interval for different time lags

Table 7-4. Input structures of machine learning models

Models	Lower prediction interval	Upper prediction interval
V_{MCS}	$RE_{t-8a}, Q_{t-1}, \Delta Q_{t-2}$	$RE_{t-0a}, Q_{t-1}, \Delta Q_{t-2}$
V_{GLUE}	$RE_{t-9a}, Q_{t-1}, \Delta Q_{t-2}$	$RE_{t-0a}, Q_{t-1}, \Delta Q_{t-2}$
V_{MCMC}	$RE_{t-8a}, Q_{t-1}, \Delta Q_{t-2}$	$RE_{t-1a}, Q_{t-1}, \Delta Q_{t-2}$
V_{SCEMUA}	$RE_{t-5a}, Q_{t-1}, \Delta Q_{t-2}$	$RE_{t-0a}, Q_{t-1}, \Delta Q_{t-2}$
V_{DREAM}	$RE_{t-4a}, Q_{t-1}, \Delta Q_{t-2}$	$RE_{t-1a}, Q_{t-1}, \Delta Q_{t-2}$
V_{PSO}	$RE_{t-6a}, Q_{t-1}, \Delta Q_{t-2}$	$RE_{t-1a}, Q_{t-1}, \Delta Q_{t-2}$
V_{ACCO}	$RE_{t-10a}, Q_{t-1}, \Delta Q_{t-2}$	$RE_{t-1a}, Q_{t-1}, \Delta Q_{t-2}$

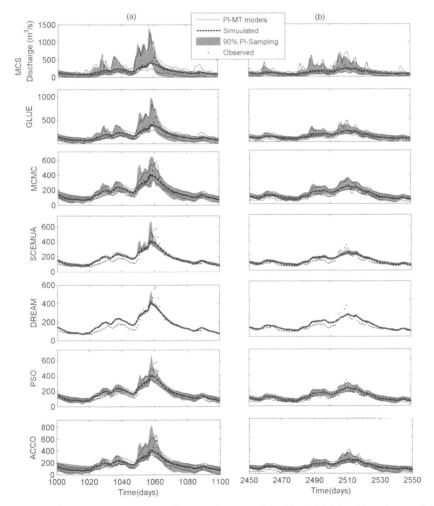

Figure 7-7. Fragment of uncertainty predicted by MT (a) calibration; and (b) verification

A multilayer perceptron neural network with one hidden layer and Levenberg-Marquardt algorithm was employed for the training in the ANN. The hyperbolic tangent function is used

135

for the hidden layer, and the linear transfer function for the output layer. The optimal number of neurons in the hidden layer, ranging from 6 to 8, was found for the lower and upper PI. The trial-and-error method was adopted to determine which, yielded the lowest CV error. Experiments with MT were carried out with pruning factor ranging from 2 to 15 numbers and for LWR models used 5 to 10 neighbours and the Tricube weight function.

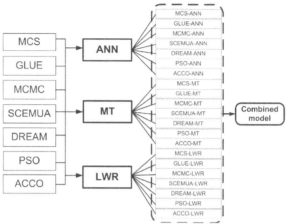

Figure 7-8. Combination of predictive uncertainty models

Table.7-5. Performances of different V models

ML Techniques	Models	CoC				RMSE			
		Lower PI		Upper PI		Lower PI		Upper PI	
		Cal	Ver	Cal	Ver	Cal	Ver	Cal	Ver
ANN	V_{MCS}	0.77	0.89	0.79	0.72	31.50	20.42	84.67	68.53
	V_{GLUE}	0.66	0.70	0.80	0.66	28.74	22.83	31.30	29.17
	V_{MCMC}	0.65	0.66	0.77	0.63	19.95	16.74	18.74	17.44
	V_{SCEMUA}	0.54	0.48	0.56	0.44	9.60	9.11	11.11	7.14
	V_{DREAM}	0.46	0.42	0.63	0.66	3.68	3.70	3.56	1.43
	V_{PSO}	0.63	0.68	0.73	0.59	15.54	12.42	14.03	11.04
	V_{ACCO}	0.65	0.66	0.73	0.59	23.18	19.60	24.46	21.70
MT	V_{MCS}	0.75	0.89	0.80	0.75	31.71	20.29	82.46	66.05
	V_{GLUE}	0.65	0.71	0.79	0.64	27.61	21.53	31.31	29.12
	V_{MCMC}	0.61	0.63	0.80	0.63	20.13	16.96	17.18	17.74
	V_{SCEMUA}	0.60	0.45	0.64	0.43	7.72	7.50	10.20	7.23
	V_{DREAM}	0.53	0.43	0.61	0.68	3.50	3.68	3.63	1.40
	V_{PSO}	0.60	0.65	0.79	0.69	15.56	12.51	9.61	8.56
	V_{ACCO}	0.62	0.65	0.76	0.58	23.24	19.22	22.99	21.85
LWR	V_{MCS}	0.91	0.85	0.93	0.74	20.63	23.35	53.13	68.64
	V_{GLUE}	0.88	0.68	0.92	0.63	17.95	23.22	21.38	29.81
	V_{MCMC}	0.87	0.61	0.91	0.60	12.81	18.03	12.01	18.61
	V_{SCEMUA}	0.86	0.41	0.86	0.42	5.05	7.84	7.44	7.34
	V_{DREAM}	0.84	0.41	0.83	0.66	2.33	3.80	2.77	1.47
	V_{PSO}	0.87	0.59	0.91	0.67	9.81	13.88	6.73	8.93
	V_{ACCO}	0.86	0.61	0.90	0.56	15.36	20.98	15.73	22.68

Table.7-6. Uncertainty indices of different V models

ML techniques	Models	PICP		MPI	
		Cal	Ver	Cal	Ver
ANN	V_{MCS}	96.94	96.79	185.81	156.86
	V_{GLUE}	87.15	82.58	115.08	100.19
	V_{MCMC}	80.29	74.02	78.25	69.88
	V_{SCEMUA}	31.14	24.77	23.09	21.07
	V_{DREAM}	9.17	6.85	7.19	6.54
	V_{PSO}	58.66	50.96	51.95	47.65
	V_{ACCO}	86.35	83.73	93.92	84.68
MT	V_{MCS}	85.56	82.30	180.59	153.94
	V_{GLUE}	88.90	82.51	110.51	98.95
	V_{MCMC}	75.54	69.16	76.84	70.76
	V_{SCEMUA}	24.75	19.49	18.99	16.83
	V_{DREAM}	9.51	7.21	7.02	6.66
	V_{PSO}	70.78	61.31	60.26	56.24
	V_{ACCO}	88.17	81.94	92.73	86.04
LWR	V_{MCS}	85.56	82.16	179.20	155.42
	V_{GLUE}	86.30	82.30	109.58	98.19
	V_{MCMC}	75.59	68.52	76.25	70.27
	V_{SCEMUA}	23.73	18.27	18.79	16.33
	V_{DREAM}	9.57	7.71	7.01	6.64
	V_{PSO}	71.18	60.96	60.08	55.39
	V_{ACCO}	84.94	80.73	92.51	85.23
SA		**76.78**	**69.81**	**78.02**	**69.00**
BMA		**83.01**	**77.80**	**98.04**	**85.68**

7.4.1 Committee of uncertainty prediction models

The process of combining a number of uncertainties predicted by machine learning models is presented in Figure 7-8. The 21 predictive uncertainty models can be built by machine learning models for the prediction of different sampling-based uncertainty BMA, and the SA method was used for combining all models. The outputs of models are multiplied by the weights from the contributions of each single model and depend on the performance of the models. The results of the BMA and SA are compared with the average PI and generated seven sampling schemes because each sampling scheme produces a different band of uncertainty, and this is difficult to compare effectively among sampling schemes.

7.4.2 Results and discussion

The results of applying the three machine learning models (ANN, MT and LWR to predict 90% PI of the HBV hydrological model outputs) showed that the percentage of the observation discharge data, falling within the prediction bounds, generated by some sampling schemes, has much lower than the given certainty level that used to produce these prediction bounds.

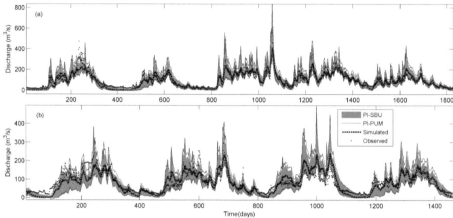

Figure 7-9. Comparison of the BMA results with average uncertainty bound (average PI) obtained from 7 PIs for the period of (a) calibration; and (b) verification

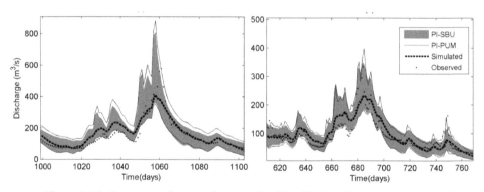

Figure 7-10. Fragment of comparison result of the BMA with mean uncertainty bound (average of seven PIs) in calibration and verification. PI-SBU is the mean PI of seven sampling-based uncertainty; PI-PUM is PI produced by using BMA of models V'.

Machine learning-based uncertainty model V trained on the data set, which was generated by each sampling scheme and tested on verification data set. Their performances are shown in Table.7-5. Table.7-6 shows the hydrograph of the 90% uncertainty bounds predicted by model V together with the mean of seven uncertainty bounds in the verification period. It can be said that the BMA reproduced the average uncertainty bounds reasonably well, in spite of the low correlation of the input variables with the PIs. Although some errors can be observed, the predicted uncertainty bounds follow the trend of average uncertainty bounds (sampling schemes). Noticeably, the mean of the uncertainty bound (sampling schemes) failed to capture the observed flow during one of the peak events (Figure 7-10a). However, the results of V (BMA) are visually covered by the observed data.

Detailed analysis reveals that the estimated uncertainty bounds from BMA contain 77.80% (*PICP*) of the observed runoffs, which is higher than the result of SA (69.81 %). The average width of the prediction intervals (*MPI*), estimated by the BMA, is wider (85.68 m³/s) than the

SA (69.00 m^3/s). Furthermore, the results showed that 15.94% of the observed data are below the lower uncertainty bounds, whereas 6.61% of the data are above the upper bounds. For the predictive capability of the BMA model in estimating lower and upper PIs, both for the calibration and verification periods, it appears that the correlation coefficient and *RMSE* for PI^L is higher than those of PI^U. This can be explained by the fact that PI^U corresponds to the higher values of flow (where the HBV model is less accurate) and has higher variability, which makes its prediction difficult.

7.5 Summary

The comparison of various predictive uncertainty models is not straightforward, so several models' outputs can be combined. A committee of predictive uncertainty models by BMA overcome the problem by conditioning, not on the single best model, but on the entire group of models (e.g.,Raftery et al., 2005). This chapter demonstrates the use of one of the methods of model averaging (however others can be employed as well), which can be employed to combine several predictive uncertainty models. BMA assumes the *pdf* of individual models of prediction in establishing the aforementioned, and uses a calibration period to determine static weights for each individual model. We have found that combination of different machine learning-based predictive uncertainty models leads to an increase in accuracy.

Chapter 8
Integration of hydrological and hydrodynamic models and their uncertainty in inundation modelling

This chapter presents an uncertainty estimation of flood inundation extent through the integration of SWAT (hydrological) and SOBEK (hydrodynamic) models. Models are implemented with MATLAB using parallel computing tools to simulate the flood inundation in the Nzoia catchment. The SWAT model simulates streamflows in the outlet of the catchment, while the SOBEK model routed the down streamflows at the outlet of the catchment to Lake Victoria. Both the SWAT and SOBEK models are individually calibrated before the coupling. Monte Carlo simulations are used to run the coupled SWAT and SOBEK, in order to produce a probabilistic flood map. Several simulations are run to explore the distribution of streamflows at the upstream (inflow boundary) of the channel (river). The resulting flood inundation map allows for presenting and analysing the probability of flooding in certain areas. Various levels of probability of flood inundation extent are identified.[6]

8.1 Introduction

Hydrological and hydraulic models simulate the river processes to assess the spatial and temporal information on flooding, which is crucial for operational flood forecasting. Hydrological models involve an empirical relation of rainfall-runoff, which reproduces the flow in the river. However, these models do not allow capturing the interaction between channel, floodplain, and backwater effects of flow in the river. Therefore, river hydraulic models (hydrodynamic models) operate to manage, adequately, the dynamic characteristics of flow by hydraulic routing. In fact, River hydraulic models are mainly used to understand the mechanisms that cause flooding, the behaviour of rivers, and the consequences of changes of future discharges and water levels. They are also used to design and evaluate the impact of flood inundations, which serves as essential information for decision makers, allowing planning and mitigating measures to be taken in time.

Hydraulic models involve governing flow equations based on the conservation of mass, momentum, and energy, which are necessary to be parameterized by imposing boundary conditions. An important aspect of hydraulic models is that these models simulate the flow, which can transform into a flood inundation map to represent the flood extent, magnitude, and shape of the flood depth. However, these models require information on river flow (e.g., boundary conditions, Manning's, channel cross section and depth), observations of flood

---

[6] Kayastha, N., van Griensven, A., Solomatine, D. P. (2011). Dealing with uncertainties in remotely linked models. *Proc. OpenWater Symposium and Workshops*, UNESCO-IHE, The Netherlands

extent (topographic data), and methods of quantifying the performance of individual simulations in reproducing inundation patterns.

One way to validate the performance of a deterministic hydraulic (flood) model is by using available remotely-sensed data on the flood extent, where data used is observed, and by comparing these data with the predicted flood extent by overlapping these two events (described in Section 8.8).

The accuracy of the models can be significantly improved using topographic data with the availability of remote sensing data. However, uncertainties in topographic data limit the use of models in river networks. Numerous sources of uncertainty need to be considered during modelling of river processes, which results in uncertain model outcomes. Knowledge of the uncertainties is crucial for a meaningful interpretation of the model results, which is significant for accurate decision-making processes (Warmink et al., 2011). Understanding and quantification of individual sources of uncertainty in flood models are very complex. Addressing such frameworks requires the knowledge of each variable that affects the flood inundation process and contributes to total model uncertainty.

Uncertainty may arise from various sources in flood inundation models. Jung et al. (2012) delineated the following sources of uncertainty: (i) flows; (ii) topography and land use data; (iii) modelling type (one-dimensional vs two-dimensional); (iv) model setup and assumptions (e. g., steady state, unsteady state); (v) model parameters (e.g., Manning's roughness); (vi) lack of model calibration data (e. g., observed flood extent); and (vii) approaches of flood inundation mapping. However, flow is one of the most uncertain variables in flood inundation mapping (Pappenberger et al., 2006). Flow can be obtained in two ways: (i) direct measurements of water stage; and (ii) estimation by hydrological model. Most flood modelling practice utilizes observed hydrographs, used for boundary conditions (Montanari et al., 2009). The uncertainty of the inflow boundary condition depends on the methods used to estimate flow. Uncertainty arises from precipitation, model structure, and the model parameters of the hydrological model, and should be included when flow is used as an input to the hydraulic model. One of the major issues is the availability of measure discharge data in un-gauged catchment. Such data are quite rare. Therefore, the hydrological model is used to overcome this problem. This model generates the simulated discharge hydrograph to assess inflow boundary conditions for the hydraulic model. In this study, the hydrological model is coupled with a hydraulic model to generate the flood inundation map, where the uncertainty outputs of the hydrological model are an input to hydraulic models.

Generally, Monte Carlo simulation is employed to estimate the uncertainty of flood models. It produces an ensemble of deterministic model simulations, where each model output received an evaluation of its value or fit based on observed flood inundation extent. This allows the determination of the impact of individual model input on the flood inundation polygon and uncertainty zone at various confidence levels. Various techniques have been used to address the issue of uncertainty in flood inundation. Examples include the Bayesian forecasting system (Krzysztofowicz 1999, 2002); GLUE (Jung et al., 2012; Blazkova and Beven 2009; Yatheendradas et al., 2008; Horritt, 2005), Fuzzy logic (Pappenberger et al., 2006), and PEST (Liu et al., 2005). Pappenberger et al. (2006) showed that the sources of uncertainty of boundary conditions have a significant effect on the performance of inundation prediction. They demonstrate remote sensing-derived water stages to correct inflow of the hydraulic models. Montanari et al. (2009) introduce the approach of calibration, and sequentially update a coupled hydrologic-hydraulic model using remotely-sensed flood information. They used remote sensing data of flooded areas for correcting volume error in the hydraulic model. Neal et al. (2009) generated an ensemble of flood simulations, where discharge is estimated from a rainfall-runoff model with precipitation from numerical weather predictions.

In this chapter, we test the uncertainty of hydrological and hydraulic models, thereby focusing on quantification of the uncertainties in the model outcomes as flood inundation. The model results are obtained using a high performance computer (parallel computing) by simulating the hydrological SWAT model and the SOBEK hydrodynamic model in the Nzoia catchment, Kenya. The SWAT model estimates the flow of the river channel, and the SOBEK model obtains the flood inundation extent. Integration of SWAT and SOBEK is done in looping sequence of model outputs, so that the final output can be utilized to estimate uncertainty.

8.2   Flood models

A flood is the result of excessive water overflow in the river channel (water bodies), that is, flow exceeds the capacity of the river channel and submerges land. The occurrence and the magnitude of the flood event are traditionally established by analysis of historical hydrological data. Computer models are set once the frequency, magnitude, and shape of the hydrograph have been established. The topographic data of the river, floodplain and land are used to estimate flood depth and flood elevation.

Flood models are built on one-dimenstional (1D), two-dimensional (2D) or coupled 1D - 2D flow, based on solutions of the full or approximate forms of the shallow water equations. 1D model can be more efficient than 2D models for flow in the main river, while 2D models are usually more accurate when flow in the floodplains is significant. 1D modelling is the most common approach for simulating flow in a river channel, where water flow in the river is assumed to flow in one dominant direction, aligned with the center line of the main river channel. A 1D model solves shallow water flows in open channels, assuming that vertical acceleration is not significant, and water level in the channel cross-section is approximately horizontal. However, problems arise when the channel is embanked and water levels are different in the floodplain than in the channel. In this case, 2D models can be used to calculate water levels in the floodplain.

In coupled 1D-2D models, unidirectional representation of a river is coupled to two-dimensional representations of the floodplains, which separate into a form of the grid using 2D solver. Fully 2D models involve river channel bathymetry and floodplain topography as an integrated continuous surface. This type of model solves the water level and depth-averaged velocities in two spatial dimensions (Pender and Neelz, 2007). Fully 2D models do not assume a dominant direction for the flow. Therefore, they are suited for floodplain flows, since there is no necessity to prescribe a particular direction of flow. Limitations of 2D models involve large computing power, longer run times, and difficulty in including hydraulic structures, such as bridges and weirs. Furthermore, large data requirements could make their use prohibitive in data-scarce regions (Merwade et al., 2008).

8.3 Model integration

Flood inundation modelling is a sequential process of hydrological analysis, which is followed by hydraulic analysis and then by geospatial processing. Each model (hydrological/hydrodynamic) involves a set of procedures that are performed in sequence. An increasing understanding of the complexity of river processes can be recognized by the looping sequence of models (hydrological and hydraulic). This is accomplished by the integration of models. While it not only incorporates the knowledge of two models, it is also able to capture the overall complex flood processes in a study area. Such integration requires accessibility of data, sufficient processing power, and complex process interactions between models. In present study, integration or linking of models assumes that the output of one model will feed another. However, the models should be linked together in a meaningful way that should address the variables, scales, and resolution problems (Voinov and Cerco, 2010). A number of tools for model integration can be found in the literature, e.g., OpenMI (Gregersen et al., 2007); the Simple Model Wrapper (SMW, Castronova and Goodall, 2010); Invisible Modelling Environment (TIME, Rahman et al., 2003); and HydroPlanner (Maheepala et al., 2005).

TIME tool is used to integrate the models in the water domain, which does not support the use of non-TIME models (e.g., SWAT, HBV, SOBEK etc.). HydroPlanner combines wastewater and stormwater systems and interacts with natural water systems, as well as contaminant and nutrient flows at the city and regional scale. This tool supports models that have spatially explicit data only. SMW reduces the complexity associated with creating new, process-level components within an OpenMI-based modelling framework. However, these tools require a broader and deeper understanding of the model components that are going to be used, and hardly support a minimal range of metadata types. They are also difficult to manage. Spatial representations, visualizations of uncertainty, and the large scale of temporal data input and output are the limitations of this tool.

Models can be coupled in two different ways (Goodall et al, 2011) (see Figure 8-1): (i) tight-coupling, in which the independent models are integrated into a single modelling application by changing the code. This provides complete control over the process representations and data structures within all parts of the model, but the data structures and semantics within a module are fixed; and (ii) loosely-coupled, in which the model component developer allows the assembly of the internal algorithms, which are required to represent a particular system. Some intermediate data processes occur, which presents an opportunity to identify model inputs that are physically impossible (Voinov and Cerco (2010). However, in tight-coupled models, the results from one model directly transmit into another

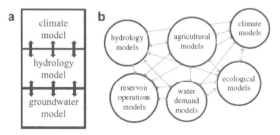

Figure 8-1. (a) tight-coupling; and (b) loose-coupling (Goodall et al. 2011).

Open Modelling Interface (OpenMI) is the result of a European-wide effort to develop a protocol making it possible to link various types of simulation models (Gregersen et al., 2005; 2007). OpenMI considerably reduced the time and effort required to import and export data between models. However, it required an adjustment of the internal time steps of models, and the selection of the appropriate type of links is needed to reduce possible numerical instabilities at selected locations (Gregersen et al., 2007). Donchyts et al. (2007) pointed out some limitations of OpenMI. Firstly, it does not explicitly support synchronization among different models at run-time to prevent one component from waiting forever, which might be inadequate for real-time applications. Secondly, there is a lack of detailed semantic information at runtime. Branger et al. (2010) mentioned that the design of OpenMI is most useful for coupling previously existing models, and it has no facilities for developing new components from scratch or establishing data input and output facilities.

8.4   Propagation of uncertainties in integrated models

The uncertainty associated with each individual procedure affects the overall system. The end-to-end (or whole system) modelling approach ensures integration of the models (e.g., climate, hydrology and hydraulics), while taking into account sources of feedbacks within the uncertainties outcomes. This type of modelling approach enables a better understanding of the complex system effects of key drivers, such as climate, in simulating how the future may unfold under various scenarios. These models produce various sources of uncertainties used to support decision-making. For example, they contribute to the analysis of the influence of climate change on the streamflows (Schaake, 1990; Xu, 1999; Chiew and McMahon, 2002; Bronstert et al., 2005; Burger et al., 2007). Coupled meteo-hydrological models are effective tools to achieve longer lead times in hydrological forecasting (Ramos et al., 2009). The cooperation between interdisciplinary domains is essential in developing such systems, as they can provide the basis for tracking uncertainty from the beginning (climate forcing data) to end (flood predictions). Pappenberger et al. (2005) investigated the cascading of model uncertainty from medium-range weather forecasts (10-days ahead of rainfall forecasts) through the LISFLOOD rainfall–runoff model down to flood inundation predictions within the European Flood Forecasting System (EFFS). They found that rainfall forecasts in the modelling system for real-time flood inundation prediction can yeild useful longer lead times for decision-making. McMillan and Brasington (2008) developed the "end-to-end" modelling approach for the formation of a coupled system of models allowing continuous simulation methodology to predict the magnitude and to simulate the effects of high return period flood events. They reported that their approaches addressed the problem of computational limitations. Saint and Murphi (2010) developed an end-to-end workflow for coupled climate and hydrological modelling to examine the effect of environmental changes. They presented data establishing that interactive systems can be used to tackle emerging questions about climate uncertainty.

Cascading uncertainty is a widely accepted technique for propagation of uncertainties through the model chain. However, it is not straightforward. Each model involved in the integrated modelling system has to be quantified for its effects. In the last few years, several approaches have been developed to manage the propagation of uncertainty in the hydro-meteorological model to the flood routing model (see Pappenberger et al., 2005; Romanowicz et al., 2006; McMillan and Brasington, 2008; Block et al., 2009; Thielen et al., 2009; He et al., 2009; Cloke and Pappenberger, 2009). Pappenberger et al. (2005a) present the cascading model uncertainty predictions and propagation of uncertainty, from an atmospheric model to a rainfall runoff model to a flood inundation model. Cloke and Pappenberger (2009) show the capturing and cascading of uncertainty through flood forecasting systems to produce an uncertainty prediction of flooding. He et al. (2009) demonstrate the coupled atmospheric-hydrologic-hydraulic cascade system, driven by the THORPEX Interactive Grand Global Ensemble (TIGGE) ensemble forecasts. A probabilistic discharge and flood inundation forecast are presented to test the use of the TIGGE database. The study shows that mainly precipitation input uncertainties mainly dominate in the system, and these propagated through the cascade chain. They suggested that the current NWPs were influenced by spatial precipitation variability that requires improvement in resolution. Techniques should be developed that narrow down the spatial gap between meteorology and hydrology.

8.5 SWAT and SOBEK models setup for the Nzoia catchment

The reach length of the longest Nzoia River channel is approximately 355 kilometres (km). Along its long path up to its outflow into the Lake Victoria, It is joined by four main tributaries on the left bank and six relatively smaller tributaries on the right bank. It has a catchment area of 12,709 square kilometers (km^2). The river originates from Mt. Elgon and Cherangani Hills. This is a highland of the catchment known to receive higher rainfall. Due to excessive rainfall, the high flows accumulating in the plains results in increased incidences of floods in the lower Nzoia catchment. The most severely affected regions are within a the Bundalangi Division of the Busia District, which is located on the shores of Lake Victoria at the mouth of the Nzoia River. The river length ranges from 0 (outfall into the lake) to 25 km, and the bed slope flattens to 1 in 3400 as the river meanders through a wide flood plain (red line shown in Figure 8-3). The channel width increases to 70 metres (m), and the height of the banks decreases considerably, which causes the spilling of flood waters over the banks, and the consequent flooding of large areas on either side. Major severe flooding was recorded in the lower reaches of the Nzoia River in November 2008, which mainly affected the Budalangi Division of the Busia District. This flooding affected more than 12,000 people.

Figure 8-2. Map of the Nzoia catchment. The hydrodynamic model SOBEK is set up in for the river reach shown in red.

The catchment hydrological simulations were undertaken using the Soil and Water Assessment Tool (SWAT), version 2005. This model is a physical-based semi-distributed model developed to simulate the processes at the catchment scale using a daily time-step (Neittsch et al., 2005). Ten years of daily data, from 1970 to 1989, and six years of daily data, from 1980 to 1985, were selected for model calibration and verification, respectively. The ACCO optimization algorithm (Solomatine 1999) is used to find the most sensitive optimal eight parameters (Table 8-1) to establish the best possible model. The surface runoff and groundwater were adjusted until a good fit was achieved between the observed and simulated streamflow. The performance of the model as the coefficient of efficiency obtained a value of 0.69 in calibration and 0.60 in verification.

Table 8-1. Ranges of SWAT parameters for model calibration

Name	Minimum	Maximum
v__Rchrg_Dp.gw	0.01	1
v__Canmx.hru	0.0	0.5
v__CH_K2.rte	20	70
v__Surlag.bsn	0.5	1.3
r__CN2.mgt	0.5	0.5
v__Gwqmn.gw	0.5	1.5
v__ALPHA_Bf.gw	0.5	1.5
r__SOL_AWC().sol	0.1	0.5

SOBEK-RE is setup for flood simulation along 18.6 km of the lower the Nzoia River (river reach mark in red in Figure 8-2). SOBEK-RE is a one-dimensional (1D)- two-dimensional (2D) hydrodynamic model that couples 1D hydraulic modelling of the river channel to a 2D representation of the floodplains. The hydrodynamic 1D-2D simulation engine is built based upon the optimum combination of a minimum connection search by direct solver and the conjugate gradient method.

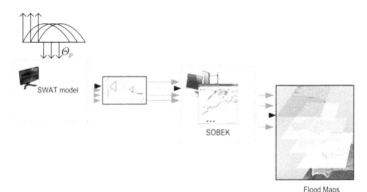

Flood Maps

Figure 8-3. Illustration for generating the multiple flood inundation map

The upstream boundary condition of the SOBEK model is set at the flow hydrograph at the hydrometric station in the SWAT model output. Topographical information from the Shuttle Radar Topography Mission (SRTM) has data with a resolution of 90 m data, used in both the hydrological and hydraulic models. The downstream boundary condition is set at the normal depth of Victoria Lake. The downstream boundary often sets the slope of the water surface in the framework of an unsteady flow analysis (Montanari et al., 2009). Three tributaries meet in the main channel between the upstream and downstream boundaries.

However, their contribution is quite small and not relevant for the flood extent information. Therefore, we did not attempt simulation.

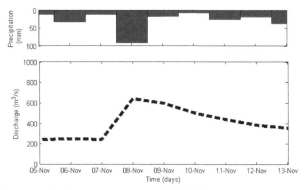

Figure 8-4. Simulated discharge generated by SWAT (05-Nov-2008 to 13-Nov-2008)

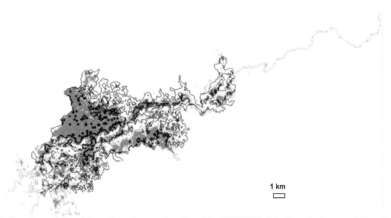

Figure 8-5 Overlay of flood-inundated area simulated by SOBEK (blue) and observed flood extent (black)

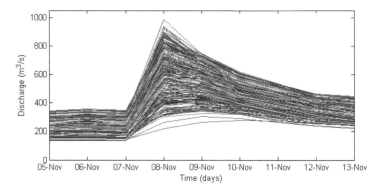

Figure 8-6. An ensemble of boundary inflow (05-Nov-2008 to 13-Nov-2008) generated by SWAT

The SOBEK model was calibrated using inflow flow data that simulated output of SWAT for the period from 6 to 13 November 2008. During the process of SOBEK model calibration, two values of Manning's for the river channel and one for the floodplain were adjusted to bridge the gap between the observed and simulated flood inundation area. The observed flood inundation area was compared to topographic information captured by the Advanced Land Imager (ALI) NASA's earth satellite on 13 November 2008 (Figure 8-11). The model set spatially uniform Manning's coefficient for the channel and flood plain, of 0.032 and 0.03, respectively. The total inundated area predicted was obtained, that is 50.16 km^2, However, the satellite image calculated 47.68 km^2. The model was run at a spatial resolution of 90 m and a time step of 20 s in order to reduce computational efficiency and random errors associated with the topographic data (DEM). The files for the inundation extent were exported using MATLAB for analysis and visualization.

The SWAT and SOBEK models were integrated after individual calibration. MC simulations are set to represent model uncertainties through an ensemble of model outputs, where each ensemble corresponds to a model realization of the set of parameters that simulate the SWAT model that produced the boundary discharges (Figure 8-3) into SOBEK. Error in boundary inflows produced uncertainty in the hydraulic model, which is estimated by an ensemble of model outputs. These are generated by treating boundary inflows in the model as stochastic variables. In this study, we represent the uncertainty in the upstream boundary inflows, which is generated by the SWAT simulations, so that the SOBEK model can produce flood inundation extent maps. The MC simulation procedure produces 25,000 ensembles of flood inundation extent maps with respect to boundary inflow hydrographs (Figure 8-6).

8.6   Approach to estimate the uncertainty of flood inundation extent

The uncertainty estimation of flood inundation extent that comes from the uncertainty of streamflows (including the inflow (upstream) boundary conditions and the hydrological model parameters) consists of the following stages:

1. Calibration and verification of the SWAT model, based on historical observations of streamflows.
2. Generation of streamflows from SWAT (inflow boundary condition) for SOBEK for particular event.
3. Generation of a priori distribution for the chosen parameters (hydrological model), using information obtained from the deterministic optimization.
4. MC simulation of the SWAT model, using sampling of a parameter space to calculate the uncertainty of streamflows.
5. Calibration of the SOBEK model using inflow boundary (simulated streamflows) that generated by SWAT, the optimal value of Mannings (parameters) in the flood plain and river channels obtained during the calibration (e.g., maximum water levels in the cross sections of the river reach and flood inundation area).
6. Multiple simulation of the SWAT and SOBEK models using parallel computing and the generation of ensembles of flood inundation maps with a specified inflow boundary condition (i.e., an event being greater or equal to a given value), corresponding to the uncertainty of the inflow boundary derived for an input of the model.

7. Estimation of uncertainty of a posteriori outputs from the models, in flood inundation maps, and the derivation of maps of probabilities of maximum inundation.

8.7 Use of parallel computing

The SWAT and SOBEK models require sets of large data and computational resources for long simulation time. When multiple simulations of models and model chains for a single computer restrict the use because of computational power and time, an alternative can be the implementation of multiple computers. This can be accomplished by distributing tasks across computers in a network, arranged in clusters, and through cloud computing with effective and efficient support from hydroinformatics tools. In this study, we show the potential of using parallel computing for analyzing uncertainty of hydrological and hydraulic systems by running multiple simulations in parallel.

The Parallel Computing Toolbox of MATLAB provides parallel application programming that allows the execution of multiple independent models simultaneously for the same problem with multiple computers, in distributed and parallel environments, by managing computations and data between MATLAB sessions and computing resources. The MATLAB Distributed Computing Server is set up on a computer cluster as workers. Models are execute parallel applications from the MATLAB prompt on these workers. Master computer retrieves results after they finish their assigned computations. The parallel processing functions allow the implement of tasks in parallel and data-parallel algorithms at a high level in MATLAB without programming for specific hardware and network architectures.

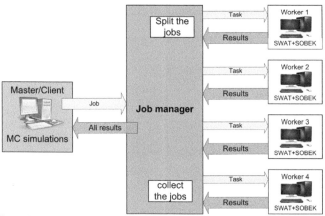

Figure 8-7. Architecture for parallelization of MC simulation studies

The architecture of parallel computing is shown in Figure 8-5, which includes several functions (scheduler, execution, retrieve, destroy) that communicates between master and workers. The optional job manager can run on any machine on the network. The scheduler creates an object in the MATLAB session to represent the job manager that will run the job. Each task of a job is represented by a task object in the local MATLAB session (worker). The task for running the job from the job manager, the worker executes this task. When workers complete the job's tasks, it moves to the finished state. The resulting data from the evaluation of the job of each task object is collected, returns the result is returned to the job manager, destroys the job to free memory resources. The worker is, then assigned another task. When all tasks for a running job have been assigned to workers, the job manager starts running the next job with the next available worker.

The flow of information through the model chain is shown in Figure 8-3. MC simulations run on a master computer, and integrated (cascade) models (SWAT and SOBEK) run on workers' computers. The output of the SWAT model results in the hydrographs at specific locations. The model results are input (i.e., boundary conditions) to the SOBEK model. The automation of data transfer and model executions were implemented by additional runtime functions and are written in MATLAB. The two functions *"changeboundary.m"*and *"changefriction.m"* are written in MATLAB scripts, these are used to change inflow boundary and friction coefficient (Manning's) in each simulation. The scripts for change boundary is shown in Figure 8-9 and for iterative runs of the model shown in Figure 8-10. After the finalization of the execution of one integrated model run for a particular input, the results are converted to text files and saved, in order to perform their uncertainty analysis following the collection of all results.

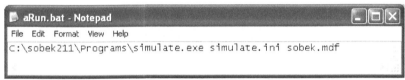

```
aRun.bat - Notepad
File   Edit   Format   View   Help
C:\sobek211\Programs\simulate.exe simulate.ini sobek.mdf
```

Figure 8-8. Batch file for runing SOBEK from the command line

```
changeboundary.m - Notepad
File   Edit   Format   View   Help
function [x]=changeboundary(bfile, prt)
% bfile='BOUNDARY.DAT';
% prt=1;
p=importdata(bfile);
pdata1=p(1:2)
pdata2=p(3:end-2) % this dat has to change
pdata3=p(end-1:end)
cvalue=[]
for lp=1:length(pdata2)
    pv= str2mat(pdata2(lp))
    pv1=pv(1:21);
%    pv2 = num2str(str2double(pv(23:26))*prt)
    pv2 = num2str(prt(lp));
    cvalue=strvcat(cvalue,[pv1 ' ' pv2,' ', '<']);
end
% cvalue=cvalue(:);
strvcat(char(pdata1),cvalue,char(pdata3))
dlmwrite('BOUNDARY.DAT', strvcat(char(pdata1),cvalue,char(pdata3)), '')
x='BOUNDARY.DAT'
```

Figure 8-9. Script for write inflow boundary data into the BOUNDARY.DAT file

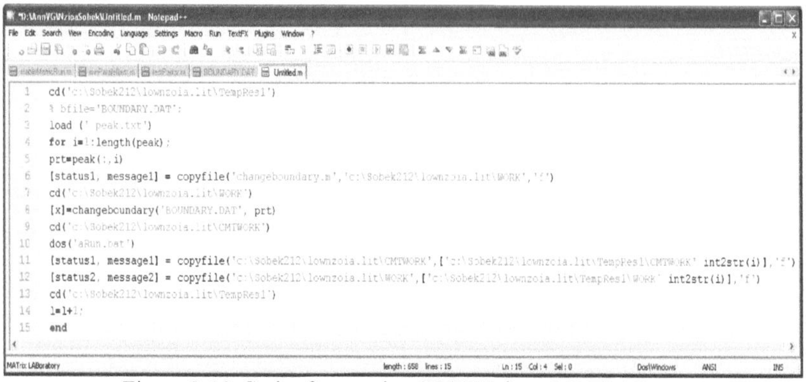

Figure 8-10. Script for running SOBEK from MATLAB

8.8 Quantification of model performance and uncertainty

The evaluation of flood models in predicting flood inundation extent depends on the types of model and observed data used. Visual representation is one of the common approaches for evaluating model outputs. However, a large data set may not permit visualization. Quantitative measures of model performance are used to compare different properties of the flood extent. Model predictions of flood inundation extent can be evaluated by retrieving flood information from space, that is, by extracting binary maps consisting of flooded and dry cells (e.g., Aronica et al., 2002; Hunter et al., 2005). The binary comparison of maps is done with the symbolic observations of measure of the fit, based on a contingency table that shows the frequency of wet and dry prediction and observation described in Table 8-2. The number of pixels correctly predicted as wet or dry, and both under-predicted and over-predicted.

The F statistic is used as a performance measure that describes the overall matching of the predicted and observed spatial distribution of flood extent. A variety of possible performance measures (F) can be found in Hunter (2005) and Schumann et al. (2009). The model we use does not cover all aspects of the inundation phenomenon, main of which are the local contribution of the rainfall, infiltration and possible backwater effects, so it is a priory expected that the inundation area will be underpredicted. Having this in mind, we have selected the F2 statistics (Schumann et al., 2009, Di Baldassarre et al., 2009) that penalizes the overprediction.

The performance measure for prediction of flood extent (Schumann et al., 2009; Di Baldassarre et al., 2009) is given below:

$$F2 = \frac{A - B}{A + B + C} \qquad (8\text{-}1)$$

where A is the size of the wet area correctly predicted by the model, *B* is the area predicted as wet that is observed to be dry (over-prediction), *C* is the wet area not predicted by the model (under-prediction), and the term −*B* in the numerator is used to penalize model over-prediction. The value of *F* ranges from −1 to +1 where value close to +1 is the model's output, comparable to the actual flood extents, while -1 exceeds the overlap size *A*.

Table 8-2. Matrix of possible flood observation in model combinations for a binary classification scheme

	Present in Observation	Absent in Observation
Present in model	A	B
Absent in model	C	D

Uncertainty in flood extent is derived by multiple simulations of a flood inundation model, where the derivation of a "probability" map can be characterized by a relative confidence measure (RCM, Romanowicz et al., 1996). This expresses a belief that uncertain prediction is a consistent representation of system behavior. The relative confidence measure for each cell j is given by a weighted-average goodness-of-fit measure:

$$RCM_j = \frac{\sum_i L_i \cdot w_{ij}}{\sum_i L_i} \tag{8-2}$$

where L_i is the weight for each simulation i, and the simulation results for the jth model element (cell), that is w_{ij}, is 1 for wet and 0 for dry. RCM_j presents the range between 0 and 1, and reflects the likelihood of inundation at that point for uncertain forcing boundary inflows. Each simulation, i, attributed a likelihood weight L_i in the range [0, 1] according to the values of measure fit F:

$$L_i = \frac{F_i - \min(F_i)}{\max(F_i) - \min(F_i)} \tag{8-3}$$

where $\max(F_i)$ and $\min(Fi)$ are the maximum and minimum measures of fit found throughout the ensemble.

Hunter et al. (2005) evaluated the binary measures that involve observing area-based measures interpreted by degree of correspondence between a given model realization and a binary pattern observation. Pappenberger et al. (2007) provided the methods of assessment using a fuzzy global performance measure with selected discrete binary performance measures for a set of model realizations to illustrate the differences between them.

Figure 8-11. Flood map observed by satellite (captured by the Advanced Land Imager (ALI) NASA's earth satellite)

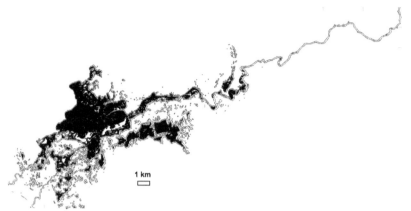

Figure 8-12. Flood inundation map observed (after processing of satellite data)

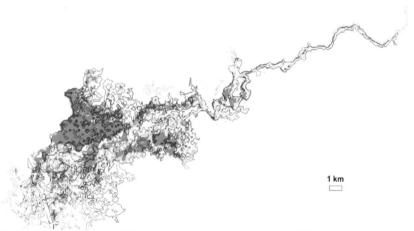

Figure 8-13. Overlay of 75% confidence level of flood inundated area (blue) and observed (black)

8.9 Results and discussion

MC simulations were conducted for uncertainty propagation from target variable inflow boundary conditions to produce different flood inundation areas. The results of 23,400 model simulations are considered for further analysis instead of the initial 25,000 MC simulations because the rest (600 simulations) less than 300 m^3/s in peak discharges (inflows boundary discharges) have no influence in flood inundation extent. Furthermore, the hydraulic model requires the specification of roughness parameters, and during MC simulations, these parameters used two Manning's as a fixed variable (used calibrated values), and the number of acceptable flood simulations were generated at without changing of Manning's. The differences between the simulated inundation areas are obtained in the range of 2.07 to 103 km^2. However, an observation satellite calculated 47.68 km^2. The satellite observed image spatial regulation 25 m (Figure 8-11) is extrapolated into 90 m (Figure 8-12) in order to simplify the analysis and make compatible comparison of the flood map obtained from

SOBEK. The inundation maps generated by SOBEK are reclassified into binary wet/dry flood maps to produce a probability of the inundated area. This represents the percentage of area in which each cell is classified as inundated. The analysis of inundation corresponding to 5 to 95% is shown in Figure 8-16.

Figure 8-14. Overlay of maximum flood inundated area simulated by SOBEK (blue) and observed (black)

The results of analysis show that the inflow boundary condition had the highest influence on the flood inundation area. The reason for the overestimated flood extent prediction could be attributed to the uncertainty of the observed topography. The importance for flood mapping is the selection of proper performance measures of models. Model prediction flood extent was evaluated against the relevant benchmark DEM simulations fit (F2) statistic (Schumann et al., 2009, Di Baldassarre et al., 2009). The likelihood measure based on the F2 produces the uncertainty bound of flood map. In Figure 8-15 one may note that the downstream part of the domain is not inundated. Figure 8-16 shows that in most simulations, the domain water enters at the blocked drain in the southwest corner and flows down the lake.

One another aspect is design flood estimation that is an additional source of uncertainty into flood inundation extent (Di Baldassarre et al., 2010). The estimated 1-in-100-year design flood is obtained in peak discharge value around 688 m3/s (Balica, 2012) for Nzoia catchment. This value is calculated based on Gumbel extreme value distribution method using a limited data set. We used this value in order to compare the results of the MC simulation to other estimates of flood magnitude (simulated by SWAT). It can observe is that the interval for the peak flood values on November 8, 2008 that found within interval [270 m3/s, 986 m3/s] indeed includes the value 688 m3/s (refer to Figure 8-6).

The effects of the non-stationarity of friction parameters can decrease performance of hydraulic models (Horritt et al., 2007), especially the modelling of flood extent. However, the probabilistic approach can be facilitated to reduce these effects because this approach uses multiple models in prediction, rather than a single best model (Bates et al., 2004).

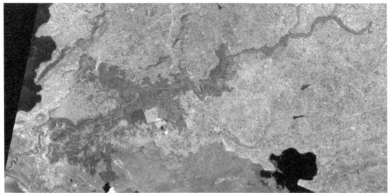

Figure 8-15. Simulated flood inundation overlapped with the Google map

Figure 8-16. Uncertainty in flood inundation map, taking account of uncertainty of model parameters
(legend shows relative confidence measure, 5% to 90%)

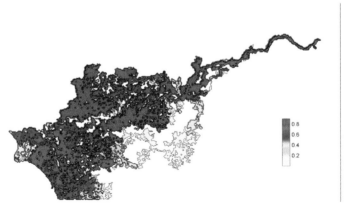

Figure 8-17. An inundation likelihood map for Nzoia (conditioning averaging of likelihood)

Figure 8-17 shows that the flood extent map, obtained by combining weights based on the likelihood of 23,400 simulations, reflects the uncertainty from inflow boundary in flood inundation extent. The higher value (after the averaging of likelihood) indicates that the model simulated flood extend is overlapped by observed flood inundation extent.

In fact, the hydrological flows are affected by errors that are caused by the inadequate model structure and inputs, as well as by the parameterization of the model itself. Therefore, it might produce a different uncertain flood extent map when uncertain sources of input and model structure take into account the uncertainty in the hydrological flow estimation. This chapter considers initial experiments with the uncertainty-based modelling, and only the hydrological model parameters are considered as the uncertainty source.

The optimal parameters for the period 1970-1985 (calibration and verification) were used in simulating the hydrological behaviour of the catchment in 2008 event. We tried to explore the magnitude of changes in Nzoia catchment. We undertook all possible efforts to find latest data from the authorities but unfortunately failed. However, the experts who know the area told us that the catchment land uses characteristics have not changed much from 1980 to 2010 since the agricultural areas largely preserve the same pattern with the main crop being sugar cane. So a conclusion was made that, as a proof of concept, we can still use the model calibrated on old data to simulate the recent event. However, if more data is collected or made available, then the model has to be re-calibrated.

8.10 Summary

Uncertainty in flood inundation extent can come from various sources of uncertainty in hydrological models. This study aimed at estimating the uncertainty of flood inundation extent downstream in the Nzoia River using the 1D -2D SOBEK river model, where uncertainty is considered from a hydrological model by simulating SWAT. MC simulations were used to generate multiple flood inundation extent patterns from coupled (cascade) SWAT and SOBEK model outputs. The uncertainty of outputs is presented in the form of confidence level, which is based on the observed flood inundation extent obtained from remote sensing data. This analysis allows one to create a visualization, indicating the uncertainty in flood inundation maps. The coupling of hydrological and hydraulic models is important for flood forecasting. However, additional tests are required for data processing techniques, which can better represent system responses especially in operational flood forecasting.

Use of the tools supporting parallel computing allows for estimating uncertainty of flood inundation extent by integrating hydrological and hydraulic models. The advantage of this framework is that it can incorporate of the hydrological uncertainty into flood inundation modelling. The disadvantage is that it is computationally more demanding and requires the use of large data sets.

Chapter 9
Conclusions and recommendations

This study aimed at refining the committee approach and uncertainty prediction in hydrological modelling. The following conclusions and recommendations are grouped around these main themes of this study: model committees, uncertainty prediction models, their committees, and uncertainty-based integrated modelling.

9.1 Committee modelling

One of the important keywords in this study is "committee". Different types of committee models, namely (a) fuzzy committee models; (b) states-based committee models; (c) inputs-based committee models; (d) outputs-based committee models; and (e) hybrid committee models, are presented in this thesis. All models used a multi-modelling approach, which intends to improve model prediction involving a combination of model outputs. These outputs are obtained by differently parameterized models with the same model structure (in the above mentioned first four models, that is, (a), (b), (c), and (d). In model (e), two different models were combined. One output was from a process-based model, and another from data-driven models. These models were combined to form a hybrid model. The resulting models were tested on verification data and compared with other models based on objective function *RMSE* and *NSE*.

The major findings from the committee modelling experiments show that the combination of two specialized models indeed does lead to the better performance of the resulting committee model. The committee model always showed better results, on both objective functions ($RMSE_{HF}$ and $RMSE_{LF}$), compared with the single model. Furthermore, fuzzy committee models always presented better than the single hydrological model with respect to independent values of parameters *MFtype* and *WStype*. One of the interesting effects here is that the membership function parameters δ and γ to be optimized often having very close values, which means that there is a very narrow region where specialized models "work together." Potentially this may which a minor change in average flows will force the committee model to produce relatively large changes in outputs. We cannot suggest a "universal" best set of parameters *MFtype* and *WStype* that would be applicable for every case study, as, in calibration, all of them were good. However, in verification performance of models using different *MFtype* and *WStype* showed slight differences depending on cases.

States and inputs-based committee models are not so advantageous in their performances when compared with fuzzy committee models. However, the outputs-based committee models have shown very good results. The states-, inputs-, and outputs-based committee models are treated by the time-varying indicators that weigh the streamflow simulations of individual models. These weights add the quantity of flows into the system that might not be acceptable for process-based modelling theory, but this is not the case for fuzzy committee models.

It should be noted that the hydrological model state variables are not always possible to use in weighting schemes because they depend completely on the accessibility of a model source code (inside) to abstract these values.

Hybrid committee models have shown great potential among committee models. As demonstrated in a committee modelling study, the integrated form of the conceptual models and the ANN model as a hybrid model offer the most promising results compared with any other committee models.

Further development and application of the presented committee modelling approach is seen in exploring possible interactions between the parameters of the weighting functions and the shapes and parameters of the membership functions.

The committee model may be sensitive to the choice of these parameters, determining the shapes of the weighting and membership functions, and indeed exploring the use of more robust optimization methods for their identification to ensure higher robustness of the committee model. The committee approach helps one to understand hydrological complexity and to improve the predictive power of the model. A more accurate comparison of performances between various models can be made if the problem of overfitting is addressed (for example, by cross-validation during model calibration and stopping the optimization process earlier to ensure minimum error on cross-validation) instead of "deep" optimization of the model on calibration set and using a single validation set. However, to use a cross-validation set would require an effort in determining the optimal splitting of the data set into three rather than two sets. This study recommends that this be done in future research if the size of the data set would allow for this.

One of the topics for further consideration is improving the performance metrics for the committee models; they use the same metric of objective functions for different magnitude of flows, which originate from statistical theory. However, the nature of this metric (e.g., *RMSE*) is basically oriented towards high flows and may not be suited for low flows. Therefore, the performance measure can be acknowledged in the form of a transformed metric (e.g., transformed *RMSE*) to calibrate low flow model (van Werkhoven et al., 2009; Willems, 2009; Kollat et al., 2012).

9.2 Sampling-based uncertainty analysis techniques

The reliability of the uncertainty estimation methods strongly depends on the sampling method used, choice of likelihood measure, dimension of estimation problems (number of model parameters), size of catchment, and the quantity and quality of observation data. Exact comparisons of these methods are difficult because they are based on different philosophies and hypotheses. The common criteria for the comparisons in this study used what are uncertainty indices, objectives function, visual plots, and ease of use. We focused on estimation of uncertainty for the hydrological model outputs produced by seven different algorithms. We explored different sampling outcomes of the conceptual HBV hydrological model. Among these sampling methods, MCS and GLUE are easy to implement, due to fewer assumptions, but more computationally intensive. GLUE often requires expert judgment for an acceptance threshold because this is subjective. The subjective choices of objective functions can be influenced by the estimation of the prediction interval. PSO requires knowledge about the movement and intelligence of swarms, and ACCO requires clustering and covering theory. MCMC, SCEMUA, and DREAM estimate the parameters describing probabilistic representation of parameter uncertainty and are used within the classical Bayesian framework.

SCEMUA and DREAM produced narrow uncertainty intervals, while MCS and GLUE produced an ensemble of models spread all over the parameter space. PSO and MCMC produced less-wide uncertainty than MCS and GLUE, but ACCO had a shaped narrower uncertainty than GLUE. The GLUE method rejects the larger number of samples when sampling was carried out with the wider ranges of the parameter space, while less samples are rejected in narrow ranges. The uncertainty statistics (e.g., prediction interval, *pdf*) is strongly influenced by the amount of samples used (generated model outputs). Therefore, a sufficient sample should accomplish to enable comprehensive evaluation of uncertainty.

MCMC, SCEMUA, DREAM, PSO and ACCO simultaneously estimate the global exploration of optimum parameter. SCEMUA, DREAM, PSO, and ACCO were originally developed for optimization, while MCS, GLUE and MCMC are explicitly oriented for uncertainty analysis. Different methods used different sampling strategies, and there are no generally applicable rules able to indicate if "unconventional" sampling adopted in the randomized search algorithms, as in DREAM, SCEMUA and ACCO, can be employed to judge model uncertainty.

The uncertainty bounds (prediction intervals) of model outputs depend on the range of parameters considered for uncertainty analysis. The narrow range of parameters represented relatively narrow uncertainty, but not always. This depends on the magnitude of selected ranges that is narrowing around the optimum parameters producing a narrow uncertainty bound.

The wider prediction interval generated by MCS (compared to that of DREAM) does not necessarily indicate that MCS is bad (visa versa, this may be a more realistic estimate of uncertainty). Different algorithms generate different number of samples in different regions of the parameter space (same for SCEMUA). This brings up a more general question to which we still do not have an answer: if unconventional (economic) different sampling strategies (like those employed in the optimization algorithms SCEMUA and ACCO) were used, how reliable would the results of uncertainty analysis be?

The prediction intervals presented by some algorithms were very narrow, and only a few observation points were covered by prediction intervals. Hence, model parameter uncertainty may be less significant than the uncertainties caused by other sources for that sampling algorithm. In such a case, it is worth considering the uncertainty associated with model structure and input so that the total uncertainty became a wider prediction interval. On the other hand, parameter variability alone could be compensated by the other sources of uncertainty if in a wider uncertainty bound.

Further research may be inspired by the present study, since the problem here is that different sampling algorithms produce different results, and there is no clear evidence to suggest that one algorithm is superior. A solution could be to form a committee of all model uncertainty and to develop a model averaging scheme to generate the final prediction interval. This would allow for the combination of different uncertainty methods, where incorporated knowledge uncertainty is an uncertain hydrologic domain.

At the same time, if enough computational resources were available, an obvious answer to the question about the sampling algorithms would be: do not use "economic" strategies trying to move towards the optimal model too quickly, or "structured" sampling like LHS, but apply a straightforward MC sampling from the prior parameters distributions, preferably from their joint distribution if available.

9.3   Uncertainty prediction using machine learning techniques

One of the issues related to uncertainty-based modelling is prediction of uncertainty (as opposed to its analysis). In this thesis, we explore the application of an approach (MLUE) to encapsulate the results of MC simulations using machine learning techniques. A machine learning model (we employed ANN, MT and LWR) were first trained on the data generated by MC simulations to encapsulate the relationship between the hydrometeorological variables and the characteristics of the model output probability distribution (prediction interval and *pdf*). Then trained models were used to estimate the prediction interval for the new input data. MC simulations were performed off-line, only to generate the data to train the model, while the trained models were employed to estimate the uncertainty (*pdf* quantiles) in real-time application without running the MC simulations further.

This method is computationally efficient and can be used in real-time application when a large number of model runs are required, and it is applicable to hydrological models. First, we tested two separate machine learning models to estimate the two quantiles (5% and 95%) forming the 90% prediction interval. Furthermore, this method extended to predict several quantiles of the model outputs, that is, in fact, to estimate the probability distribution of the model output generated by MC simulations. The results demonstrate that the method performs quite well in estimating the *CDF*, resulting from the MC simulations. Several performance indicators and visual inspection show that machine learning models are reasonably accurate to approximate the uncertainty bounds. It is also observed that the uncertainty bounds estimated by ANN, MT and LWR are comparable. However, ANN is somewhat better than the other two models.

The results demonstrate that the prediction of uncertainty with machine learning techniques generate interpretable uncertainty estimates and quite accurate, and this is an indicator that the presented method can be a valuable tool for the assessment of the uncertainty of various predictive models. Furthermore, it allow for assessing uncertainty of complex models in real time.

Applicability of the MLUE approach in any sampling methods can be recommended, ensuring the compatibility of the models for multiple quantiles to achieve monotonicity of the resulting approximation of *CDF*, considering multiple sources of uncertainty, and testing the method on more complex models.

Further studies should aim at testing the applicability of this approach with other sampling methods (e.g., MCMC, SCEMUA, DREAM, and others), ensuring compatibility of the models for multiple quantiles to achieve monotonicity of the resulting approximation of *CDF*, considering multiple sources of uncertainty, and testing the method on more complex models. Furthermore, future studies are intended to test other machine learning models, and to apply the presented methodology to other hydrological (process) models in various case studies.

9.4   Committee of predictive uncertainty models

The two mentioned keywords, "committee" and "uncertainty", come here together. Machine learning techniques use the results of MC sampling (or any other sampling scheme) to build machine learning models and these models able to predict uncertainty (quantiles) of the deterministic outputs from hydrological models (described in Chapter 6). We present the results of hydrological model output uncertainties predicted from a number of machine learning models (ANN, MT and LWR). Firstly, a combination of six different model input structures is tested to predict the uncertainty of streamflow simulation from a conceptual hydrological model (HBV for the Bagmati catchment in Nepal). Several input datasets used to train predictive uncertainty models resulted in a total of 18 models. Secondly, uncertainty outputs that were generated by seven sampling methods, namely MCS, GLUE, MCMC, SCEMUA, DREAM, PSO and ACCO, used to train models, led to several predictive uncertainty models. Three machine learning models predicting seven sampling outcome uncertainties resulted in 21 models, applied to estimate the uncertainty of streamflows simulation from a conceptual hydrological model in the Nzoia catchment in Kenya.

It may not be fair to compare the results of various predictive uncertainty models. Model averaging overcomes the problem by conditioning, not on a single best model, but on the entire group of models. We present one of the methods for model averaging that can be employed to combine several predictive uncertainty models. BMA is proposed to form a committee of all predictive uncertainty models and to generate the final output. In two case studies, we demonstrate the BMA method for combining each quantile (5% and 95%), forming the lower and upper PIs (distance from the simulated output to the selected quantile). Several performance indicators and visual inspection show that the BMA of machine learning models is reasonably accurate. The verification results show that both averaging methods (BMA and SA) generally improve the predictive performance, but BMA is somewhat better than SA. Future studies intend to test other model averaging methods (dynamic averaging) in various case studies.

9.5   Uncertainty analysis of flood inundation models

The description of flood processes and their spatial representation can be realized by integrating hydrological and hydraulic models. Properly describing the uncertainties is a challenge in the linked modelling system (integration of the SWAT (hydrological) and the SOBEK (hydrodynamic) model) because of the multiple sources of uncertainty.

This study aimed at estimating uncertainty concerning potential flood inundation downstream of the Nzoia River, using 1D -2D SOBEK river model. MC simulations were used to run on coupled SWAT and SOBEK to generate multiple flood inundation extents and to produce probabilistic flood maps. Different levels of probability of flood inundation were identified in the form of percentage in confidence level, which is based on observed patterns of flood inundation obtained from remote sensing data. Uncertainty in flood inundation mapping was considered from sources of uncertainty from hydrological modelling.

The coupling of hydrological and hydraulic models with remote sensing data can be established as a powerful approach in particular for flood forecasting. However, more tests are required for different ways of both model evaluation and remote sensing data processing techniques, which can represent better system responses, especially in operational flood forecasting.

Advances in computer technology (we used MATLAB toolboxes) have led to improvements in the overall simulation of coupled hydrological and hydraulic models and allow an estimate of uncertainty for flood inundation extent. It should be noted that it is still computationally demanding.

Further development and application of the presented approach may lie in exploring various sources of uncertainty and their interactions, and the sensitivity of the resulting probabilistic maps to these sources.

9.6   Final conclusion

Overall, this thesis presents research efforts in: (i) committee modelling of hydrological models, (ii) hybrid committee hydrological models, (iii) influence of sampling strategies on prediction uncertainty of hydrological models, (iv) uncertainty prediction using machine learning techniques, (v) committee of predictive uncertainty models and (vi) flood inundation model and their uncertainty.

The main objectives and research questions have been addressed and mainly answered; the main results have been published in peer-reviewed journals. At the same time, the future suggestions for subsequent research have been outlined as well.

This study may allow for advancing the theory and practice of hydrological and integrated modelling. The developed software is made available for public use and can be used by the researchers and practitioners to advance the mentioned areas further.

This study is a contribution to Hydroinformatics, which aims to connect various scientific disciplines: hydrological modelling, hydrodynamic modelling, multi-model averaging, machine learning and data driven models, hybrid hydrological models, uncertainty analysis and high performance computing.

References

Abrahart, R. J. and See, L. M. (2002). Multi-model data fusion for river flow forecasting: an evaluation of six alternative methods based on two contrasting catchments, *Hydrology and Earth System Sciences*, 6, 655-670.

Aha, D., Kibler, D., Albert, M. (1991). Instance-based learning algorithms, *Machine Learning*, 6, 1, 37-66.

Ajami, N. K., Duan, Q., Gao, X. and Sorooshian, S. (2006). Multi-model combination techniques for hydrological forecasting: Application to Distributed Model Inter-comparison Project results, *Journal of Hydrometeorology*, 7, 755-768.

Ajami, N. K., Duan, Q., Sorooshian, S. (2007). An integrated hydrologic Bayesian multimodel combination framework: confronting input, parameter, and model structural uncertainty in hydrologic prediction, *Water Resources Research*, 43, W01403.

Allen, G. R., Pereisa, S. A., Raes, D. and Smith, M. (1998). Crop evapotranspiration, Guideline for computing crop water requirements-FAO, *Irrigation and drainage paper*, 56.

Anctil, F., Tape, D. G., (2004). An exploration of artificial neural network rainfall-runoff forecasting combined with wavelet decomposition, *Journal of Environmental Engineering Science*, 3, 121-128.

Arnold, J. G., Srinivasan, R., Muttiah, R. S., Williams, J. R. (1998). Large area hydrologic modeling and assessment Part I: model development, *Journal of American Water Resources Association*, 34, 1, 73-89.

Arnold, J. G. and Fohrer, N. (2005). Current capabilities and research opportunities in applied watershed modeling, *Hydrological Processes*, 19, 563-572.

Aronica, G., Bates, P. D. and Horritt, M. S. (2002). Assessing the uncertainty in distributed model predictions using observed binary pattern information within GLUE, *Hydrological Processes*, 16 ,10, 2001-2016.

Atiquzzaman, M., Liong, S. and Yu, X. (2006). Alternate Decision Making in Water Distribution Network with NSGA-II, *Journal of Water Resource Planning and Management*, 132, 2, 122-126.

Ballio, F. and Guadagnini, A. (2004). Convergence assessment of numerical Monte Carlo simulations in groundwater hydrology, *Water Resources Research*, 40, W04603.

Bardossy, A. and Singh, S. (2008). Robust estimation of hydrological model parameters, *Hydrology and Earth System Sciences*, 12, 1273-1283.

Barreto, W., Vojinovic, Z., Price, R. and Solomatine, D. (2010). Multiobjective Evolutionary Approach to Rehabilitation of Urban Drainage Systems, *Water Resources Research Planning and Management*, 136, 5.

Barreto, W., Vojinovic, Z., Price, R. K., Solomatine, D. P. (2009). A multi-criteria platform for the rehabilitation of urban drainage systems, Proceedings 8th International Conference on Hydroinformatics, Conception, Chile, January 2009.

Bates, B. C., and Campbell, E. P. (2001). A Markov chain Monte Carlo scheme for parameter estimation and inference in conceptual rainfall-runoff modeling, *Water Resources Research*, 37, 4, 937-948.

Bates, P. D., Horritt, M. S., Aronica, G. and Beven, K. (2004). Bayesian updating of flood inundation likelihoods conditioned on flood extent data, *Hydrological. Processes*, 18, 3347-3370.

Becker, A. and Kundzewicz, Z.W. (1987). Nonlinear flood routing with multilinear models, *Water Resources Research*, 23, 1043-1048.

Bekele, E. G. and Nicklow, J. (2007). Objective automatic calibration of SWAT using NSGA-II, *Journal of Hydrology*, 341, 165-176.

Balica, S. (2012) Applying the Flood Vulnerability Index as a Knowledge base for flood risk assessment, PhD thesis, UNESCO-IHE, Delft

Bell, V. A. and Moore, R. J. (2000). The sensitivity of catchment runoff models to rainfall data at different spatial scales, *Hydrology and Earth System Sciences*, 4, 4, 653-667.

Bergström, S. (1976). Development and application of a conceptual runoff model for Scandinavian catchments. SMHI Reports RHO, No. 7, Norrköping, Sweden.

Bergström, S. and Forsman, A. (1973). Development of a conceptual deterministic rainfall-runoff model, *Nordic Hydrology*, 4, 147-170.

Beven, K. J. (2009). Environmental modelling: an uncertain future? : an introduction to techniques for uncertainty estimation in environmental prediction, London ; New York.

Beven, K J. (2000). Uniqueness of place and process representations in hydrological modeling, *Hydrology and Earth System Sciences*, 4, 203-213.

Beven, K. and Freer, J. (2001). Equifinality, data assimilation, and uncertainty estimation in mechanistic modelling of complex environmental systems using the GLUE methodology, *Journal of Hydrology*, 249, 11-29.

Beven, K. and Kirkby, M. (1979). A physically based, variable contributing area model of basin hydrology, *Hydrological Sciences Bulletin*, 24, 1, 43-69.

Beven, K. J. (2006). A manifesto for the equifinality thesis, *Journal of Hydrology*, 320, 1-2, 18-36.

Beven, K. and Binley, A. (1992). The future of distributed models: Model calibration and uncertainty in prediction, *Hydrological Processes*, 6, 279-298.

Blasone, R., Madsen, H. and Rosbjerg D. (2007). Parameter estimation in distributed hydrological modelling: comparison of global and local optimisation techniques, *Nordic Hydrology*, 38, 4-5, 451-476.

Blasone, R., Madsen, H., Rosbjerg, D. (2008b). Uncertainty assessment of integrated distributed hydrological models using GLUE with Markov chain Monte Carlo sampling. *Journal of Hydrology*, 353, 18-32.

Blasone, R., Vrugt, J., Madsen, H., Rosbjerg, D., Robinson, B., Zyvoloski, G. (2008a). Generalized likelihood uncertainty estimation (GLUE) using adaptive Markov Chain Monte Carlo sampling, *Advances in Water Resources*, 31, 630-648.

Blazkova, S., and Beven, K. (2009). A limits of acceptability approach to model evaluation and uncertainty estimation in flood frequency estimation by continuous simulation: Skalka catchment, Czech Republic,*Water Resources Research* ,45, W00B16.

Block, P. J., Souza Filho, F. A., Sun, L., Kwon, H. H. (2009). A streamflow forecasting framework using multiple climate and hydrological models, *Journal of the American Water Resources Association*, 45, 4, 828-843.

Boucher, M.A., Lalibert´e, J. P. and Anctil, F. (2010). An experiment on the evolution of an ensemble of neural networks for streamflow forecasting, *Hydrology and Earth System Science*, 14, 603-612.

Bowden, G. J., Dandy, G. C. and Maier, H. R. (2005). Input determination for neural network models in water resources applications, *Journal of Hydrology*,301, 1-4, 75-92.

Box, G. E. P., and Tiao, G. C. (1973). Bayesian in ference in statistical analysis, Addison- Wesley-Longman, Reading, MA.

Box, G. E. P. and Jenkins, G. M. (1970). *Time series analysis*. Holden-Day, San Francisco, CA, USA.

Branger, F., Braud, I., Debionne, S., Viallet, P., Dehotin, J., Henine, H., Nedelec, Y. Anquetin, S.(2010). Towards multi-scale integrated hydrological models using the LIQUID framework. Overview of the concepts and first application examples, *Environmental Modelling and Software*, 25, 1672-1681.

Braun, L.N. and Renner, C.B. (1992). Application of a conceptual runoff model in different physiographic regions of Switzerland, *Hydrological Sciences Journal*, 37, 3, 217-232.

Breiman, L. (1996). Bagging predictors, *Machine Learning*, 24,2, 123-140.

Breiman, L., Friedman, J., Olshen, R. and Stone, C. (1984). *Classification and regression trees*. Chapman & Hall/CRC, Boca Raton, FL, USA.

Bronstert, A., Carrera, J., Kabat, P., Lütkemeier, S. (Eds.), (2005). *Coupled Models for the Hydrological Cycle – Integrating Atmosphere, Biosphere and Pedosphere*. Springer, New York, 348.

Bruen, M. (1985). Black box modelling of catchment behaviour. Ph.D. thesis, National University of Ireland, Dublin.

166

Burger, C. M., Kolditz, O., Fowler, H. J., Blenkinsop, S. (2007). Future climate scenarios and rainfall runoff modelling in the Upper Gallego catchment (Spain), Environmental Pollution 148, 842-854.

Burnash, R. J. C., Ferral, R. L. and McGuire, R. A. (1973). *A generalized streamflow simulation system, conceptual modeling for digital computers.* Report by the Joint Federal State River Forecasting Centre, Sacramento, CA, USA.

Castronova, A. M., Goodall J. L. (2010). A generic approach for developing process-level hydrologic modeling components, *Environmental Modelling and Software*, 25, 819-825.

Cavadias, G. and Morin, G. (1986). The Combination of Simulated Discharges of Hydrological Models, Application to the WMO Intercomparison of Conceptual Models of Snowmelt Runoff, *Nordic Hydrology*, 17, 1, 21-32.

Chalise, S. R., Shrestha, M. L., Thapa, K. B., Shrestha, B. R. and Bajracharya, B. (1996), Climatic and Hydrological Atlas of Nepal, *International Centre for Integrated Mountain. Development*, Kathmandu, Nepal.

Chen, J. and Adams, B. J. (2006). Integration of artificial neural networks with conceptual models in rainfall-runoff modeling, *Journal of Hydrology*, 318, 232-249.

Chen, R.S., Pi, L.C., and Hsieh, C.C. (2005). A Study on Automatic Calibration of Parameters in Tank Model. *Journal of the American Water Resources Association*, 389-402.

Chiew, F. H., & McMahon, T. A. (2002). Modelling the impacts of climate change on Australian streamflow, *Hydrological Processes*, 16, 6, 1235-1245.

Chou C. M. (2012). Particle Swarm Optimization for Identifying Rainfall-Runoff Relationships, *Journal of Water Resource and Protection*, 4, 115-126.

Chu, H. J. and Chang, L. C. (2009). Applying particle swarm optimization to parameter estimation of the nonlinear Muskingum model, *Journal of Hydrologic Engineering*, ASCE, 14, 9, 1024-1027.

Clarke, R. (1973). A review of some mathematical models used in hydrology, with observations on their calibration and use, *Journal of Hydrology*, 19, 1, 1-20.

Cleveland, W.S. and Loader, C. (1994). *Smoothing by local regression: principles and methods*, AT and T Bell Laboratories, Statistics Department, Murray Hill, NJ, USA

Cloke, H. L. and Pappenberger, F. (2009). Ensemble flood forecasting: a review, *Journal of Hydrology*, 375, 3-4, 613-626.

Corzo, G. and Solomatine, D. P. (2007a). Baseflow separation techniques for modular artificial neural network modelling in flow forecasting, *Hydrological Science Journal*, 52, 491-507.

Corzo, G. and Solomatine, D. P. (2007b). Knowledge-based modularization and global optimization of artificial neural network models in hydrological forecasting, Neural Networks 20, 528-536.

Corzo, G., Solomatine, D., Hidayat, de Wit, M., Werner, M., Uhlenbrook, S. and Price, R.K. (2009). Combining semi-distributed process-based and data-driven models in flow simulation: a case study of the Meuse river basin. *Hydrology and Earth System Sciences*, 6, 1, 729-766.

Coulibaly, P., Hache, M., Fortin, V., Bobe´e, B. (2005). Improving Daily reservoir inflow forecasts with model combination, ASCE *Journal of Hydrologic Engineering*, 10, 2, 91-99.

Crawford, N. and Linsley, R. (1966). *Digital simulation in hydrology: Stanford Watershed Model IV.* Technical Report No. 39, Department of Civil Engineering, Stanford University, Stanford, CA, USA.

Cullmann J, Krausse T. and Saile , P. (2011). Parameterising hydrological models – Comparing optimisation and robust parameter estimation, *Journal of Hydrology*, 404, 323–331.

Cullmann J., Krauße T., Philipp A. (2008). Enhancing flood forecasting with the help of processed based calibration, *Physics and Chemistry of the Earth*, 3, 1111–1116.

Dawson, C.W. and Wilby, R. (1998). An artificial neural network approach to rainfall-runoff modelling. *Hydrological Science Journal*, 43, 1, 47-66.

Deb, K., Pratap, A., Agarwal, S. and Meyarivan, T. (2002). A Fast and Elitist Multiobjective Genetic

Algorithm: NSGA-II. IEEE, *Transactions on Evolutionary Computation*, 6, 2, 182-197.

Devineni, N., Sankarasubramanian, A. and Ghosh S. (2008). Multimodel ensembles of streamflow forecasts: Role of predictor state in developing optimal combinations, *Water Resources Research*, 44.

DHM (1998). *Hydrological records of Nepal, stream flow summary*, Department of Hydrology and Meteorology, Kathmandu, Nepal.

Di Baldassarre, G., Schumann, G., and Bates, P. D. (2009). A technique for the calibration of hydraulic models using uncertain satellite observations of flood extent, *Journal of Hydrology*, 367,3, 276-282.

Di Baldassarre, G., Schumann, G., Bates, P. D., Freer, J. E. and Beven, K. J. (2010). Flood-plain mapping: a critical discussion of critical discussion of deterministic and probabilistic approaches. *Hydrological Science Journal*, 55, 3, 364-376.

Donchyts, G., Treebushny, D., Primachenko, A., Shlyahtun, N.,and Zheleznyak, M.,(2007). The architecture and prototype implementation of the Model Environment system, *Hydrology and Earth System Sciences Discussion*, 4, 75-89, 2007.

Dotto, C. B. S., Mannina, G., Kleidorfer, M., Vezzaro, L., Henrichs, M., McCarthy, D. T., Freni, G., Rauch, W. and Deletic, A. (2012). Comparison of different uncertainty techniques in urban stormwater quantity and quality modelling, *Water Research*, 46, 8, 2545-2558.

Duan Q., Ajami N. K., Gao, X. and Sorooshian, S. (2007). Multi-model ensemble hydrologic prediction using Bayesian model averaging, *Advances in Water Resources*, 30, 5, 1371-1386.

Duan, Q., Sorooshian, S., Gupta, V. (1992). Effective and efficient global optimization for conceptual rainfall-runoff models, *Water Resources Research*, 28, 1015-1031.

Dumedah, G., Berg., A. A., Wineberg, M. and Collier, R. (2010). Selecting Model Parameter Sets from a Trade-off Surface Generated from the Non-Dominated Sorting Genetic Algorithm-II, *Water Resources Management*, 24, 4469-4489.

Efstratiadis, A. and Koutsoyiannis, D. (2008). Fitting hydrological models on multiple responses using the multiobjective evolutionary annealing–simplex approach. In: Practical Hydroinformatics: Computational Intelligence and Technological Developments in Water Applications (ed. by R. J. Abrahart, L. M. See & D. P. Solomatine), 259-273. Springer Water Science and Technology Library, vol. 68, Springer-Verlag, Berlin, Germany.

Efstratiadis, A., and Koutsoyiannis, D. (2010). One decade of multi-objective calibration approaches in hydrological modelling: a review. *Hydrological Sciences Journal–Journal Des Sciences Hydrologiques*, 55, 1, 58-78.

Engeland, K., Xu, C.Y. and Gottschalk, L. (2005). Assessing uncertainties in a conceptual water balance model using Bayesian methodology, *Hydrological Sciences Journal*, 50,1, 45-63.

Fenicia, F., Savenije, H. H. G., Matgen, P., and Pfister, L. (2006). A comparison of alternative multi-objective calibration strategies for hydrological modeling, *Water Resources Research*, 43, W03434.

Fenicia, F., Solomatine, D. P., Savenije, H. H. G. and Matgen, P.(2007). Soft combination of local models in a multi-objective framework, *Hydrology and Earth System Sciences*, 11, 1797-1809.

Fernando, A., Shamseldin, A., and Abrahart, R.(2012). Use of Gene Expression Programming for Multimodel Combination of Rainfall-Runoff Models, *Journal of Hydrologic Engineering*, 17, 9, 975–985.

Feyen, L., Milan, K., and Vrugt, J. (2008). Semi-distributed parameter optimization and uncertainty assessment for large-scale streamflow simulation using global optimization, *Hydrological Sciences Journal*, 53, 2, 293-308.

Feyen, L., Vrugt, J., Ó Nualláin, B., van der Knijff, J., De Roo, A. (2007). Parameter optimization and uncertainty assessment for large scale streamflow simulation with the LISFLOOD model, *Journal of Hydrology*, 332, 276–289.

Fleming, G. (1975). *Computer simulation techniques in hydrology*. Elsevier, New York, NY, USA.

Freer, J. and Beven, K. (1996). Bayesian estimation of uncertainty in runoff prediction and the value of data: An application of the GLUE approach, *Water resource research*, 32, 7, 161-2173.

168

Freer, J., Beven, K. and Ambroise, B. (1996). Bayesian estimation of uncertainty in runoff prediction and the value of data: An application of the GLUE approach. *Water Resources Research*, 32,7, 2161-2173.

Freund, Y. and Schapire, R.E. (1996). Experiments with a new boosting algorithm. In: L. Saitta (ed.), *Processing. of 13th International Conference on Machine Learning*, Bari, Italy, Morgan Kaufmann, San Francisco, CA, USA, 148-156.

Gattke, C. and Schumann, A. (2007). Comparison of different approaches to quantify there liability of hydrological simulations , *Advance.Geoscience*, 11, 15-20.

Gelman, A., and Rubin, D. B. (1992). Inference from iterative simulation using multiple sequences. *Statistical Science*, 7,457–472.

Georgakakos, K. P., Seo, D.J., Gupta, H. V., Schaake, J., and Butts, M. B. (2004). Towards the characterization of streamflow simulation uncertainty through multimodel ensembles, *Journal of Hydrology*, 298, 222-241.

Gill, M. K., Kaheil, Y. H., Khalil, A., McKee, M., and Bastidas, L. (2006). Multiobjective particle swarm optimization for parameter estimation in hydrology. *Water Resources Research*, 42. 7.

Gilks, W. R., and Berzuini, C. (2001). Following a moving target—Monte Carlo inference for dynamic Bayesian models. *Journal of the Royal Statistical Society: Series B (Statistical Methodology)*, 63.1, 127-146.

Goldberg, D.E. (1989). Genetic algorithms in search, optimization and machine learning, *Reading: Addison-Wesley*.

Goodall, J. L., Robinson, B. F., and Castronova, A. M. (2011). Modeling water resource systems using a service-oriented computing paradigm, *Environmental Modelling and Software*, 26, 5.

Goovaerts, P.(1997), *Geostatistics for Natural Resources Evaluation*, Oxford University Press, New York (1997)

Gregersen, J. B., Gijsbers, P. J. A., Westen, S. J. P. (2007). OpenMI: open modelling interface. *Journal of Hydroinformatics*, 9, 3, 175-191.

Gregersen, J., Gijsbers, P., Westen, S., Blind, M., (2005). OpenMI: the essential concepts and their implications for legacy software, *Advances in Geosciences*, 4, 37-44.

Guo, J., Zhou, J., Song, L, Zou, Q., Zeng, X. (2012). Uncertainty assessment and optimization of hydrological model with the Shuffled Complex Evolution Metropolis algorithm:an application to artificial neural network rainfall-runoff model, *Stochastic Environmental Research and Risk Assessment*, 27, 4, 985-1004.

Gupta, H. V., Sorooshian, S., and Yapo, P. O. (1998). Toward improved calibration of hydrologic models: Multiple and noncommensurable measures of information, *Water Resources Research*, 34, 751-763.

Guyon, I., and Elisseeff, A. (2003). An Introduction to Variable and Feature Selection, *Journal of Machine Learning Research*, 3, 1157-1182.

Guzha, A. C., and Hardy, T. B., (2010). Simulating streamflow and water table depth with a coupled hydrological model, *Water Science and Engineering*, 3, 3, 241-256.

Haario, H., Laine, M., Mira, A., and Saksman, E.(2006). DRAM: Efficient adaptive MCMC, *Statistics and Computing*, 16, 339-354.

Hastings, W. K. (1970). Monte Carlo sampling methods using Markov Chains and their applications, *Biometrika*, 57, 97-109.

Haykin, S. (1999). Neural Networks: A Comprehensive Foundation, *Prentice-Hall Inc.*, Englewood Cliffs, NJ, USA.

He, H. Y., Cloke, H. L., Wetterhall, F., Pappenberger, F., Freer, J., Wilson, M. (2009). Tracking the uncertainty in flood alerts driven by grand ensemble weather predictions, *Meteorological Applications*, 16, 1, 91-101.

He, M., Hogue, T. S., Franz, K. J., Margulis, S. A., and Vrugt, J. A. (2011). Characterizing parameter sensitivity and uncertainty for a snow model across hydroclimatic regimes, *Advances in Water Resources*, 34,1, 114-127.

Herstein, L., and Filion, Y. (2011). Life-Cycle Analysis of Water Main Materials in Multi-Objective Design of Water Networks, *Journal of Hydroinformatics*,13, 3, 346-357

Horritt, M. S. (2005). Parameterisation, validation and uncertainty analysis of CFD models of fluvial and flood hydraulics in the natural environment, Chapter 9, Computational Fluid Dynamics Applications in Environmental Hydraulics, P. D. Bates, S. N. Lane and R. I. Ferguson, eds., Wiley, Chichester, UK, 193–213

Hornberger, G. M., and Spear, R. C. (1981). Approach to the preliminary analysis of environmental systems, *Journal of Environmental Management*, 12, 1.

Horritt, M. S., Di Baldassarre, G., Bates, P. D. and Brath, A. (2007). Comparing the performance of 2-D finite element and finite volume models of floodplain inundat ion using airborne SAR imagery, *Hydrological Processes*, 21, 2745-2759.

Hostache, R., Matgen, P., Montanari, A., Montanari, M., Hoffmann, L., and Pfister, L. (2011). Propagation of uncertainties in coupled hydro-meteorological forecasting systems: A stochastic approach for the assessment of the total predictive uncertainty, *Atmospheric Research*, 100, 263-274.

Hunter, N. M., Horritt, M. S., Bates, P. D., Wilson, M. D. & Werner, M. G. F. (2005). An adaptive time step solution for raster-based storag e cell modelling of floodplain inundation. *Advances in Water Resources*, 28, 9, 975-991.

Hunter, N. M. (2005). Development and assessment of dynamic storage cell codes for flood inundation modelling, Ph.D. thesis, University of Bristol, U. K.

Jacobs, R., Jordan, M., Nowlan, S. and Hinton, G. (1991). Adaptive mixtures of local experts. *Neural Computation*, 3, 1, 79-87.

Jain, A. and Srinivasulu, S. (2006). Integrated approach to model decom-posed flow hydrograph using artificial neural network and conceptual techniques, *Journal of Hydrology*, 317, 291-306.

Jeong, D. I., and Kim, Y-O. (2009). Combining single-value streamflow forecasts – A review and guidelines for selecting techniques, *Journal of Hydrology*, 377, 284–299.

Jeremiah, E., Sisson, S., Marshall, L., Mehrotra, R. and Sharma, A. (2011), Bayesian calibration and uncertainty analysis of hydrological models: A comparison of adaptive Metropolis and sequential Monte Carlo samplers, *Water Resources Research*, 47, W07547.

Jiang, Y., Liu, C., Huang, C., Wu, X. (2010). Improved particle swarm algorithm for hydrological parameter optimization, *Applied Mathematics and Computation*, 217, 3207–3215.

Jin, X., Xu, C.-Y., Zhang, Q., Singh, V. P., (2010). Parameter and modeling uncertainty simulated by GLUE and a formal Bayesian method for a conceptual hydrological model, *Journal of Hydrology*, 383 (3–4), 147-155.

Juemou, W., Ruifang, Z. and Guanwu, X. (1987). Synthesised Constrained Linear System (SCLS), *Journal of Hydraulic Engineering*, 7, Beijing.

Jung, Y and Merwade, V. (2012). Uncertainty Quantification in Flood Inundation Mapping Using Generalized Likelihood Uncertainty Estimate and Sensitivity Analysis, *Journal of Hydrologic Engineering*,17, 4, 507-520.

Karunasinghe, D. S. K., Liong, S. Y., (2006). Chaotic time series prediction with a global model: artificial neural network, *Journal of. Hydrology.*, 153, 23–52

Kayastha, N. (2006). Comparison of novel approaches of hydrological modelling, M.Sc. thesis, UNESCO-IHE, Delft, The Netherlands.

Kayastha, N. and Solomatine, D.P. (2013). Graphical materials from studies on committee modelling and uncertainty analysis report, UNESCO-IHE, Institute for Water Education, Delft, 2013, www.unesco-ihe.org/hi/sol/papers/KSGraphics.pdf

Kayastha, N., Shenlang, L., Betrie G., Zakayo Z., van Griensven A., Solomatine, D. P. (2011). Dynamic linking of the watershed model SWAT to the multi-objective optimization tool NSGAX. Proc.8th IWA Symposium on Systems Analysis and Integrated Assessment, Watermatex 2011 Spain

Kayastha, N., Ye, J., Fenicia, F., Kuzmin, V., and Solomatine, D. P. (2013). Fuzzy committees of specialized rainfall-runoff models: further enhancements and tests, Hydrology and Earth System Sciences, 17, 4441-4451.

Keating, E. H., Doherty, J., Vrugt, J. A. and Kang, Q. (2010). Optimization and uncertainty assessment of strongly nonlinear groundwater models with high parameter dimensionality, *Water Resources Research*, 46, W10517.

Keefer, T. N. and McQuivey, R. S. (1974). Multiple linearization flow routing model, *Journal of Hydraulic Division*, ASCE, 100, 7, 1031-1046.

Kennedy J. and Eberhart, R. (1995). Particle Swarm Optimization. *Proceedings of IEEE International Conference on Neural Network*, Piscataway, NJ. 1942-1948.

Khu, S. and Madsen, H. (2005). Multiobjective calibration with Pareto preference ordering: An application to rainfall-runoff model calibration, *Water Resources Research*, 41, W03004.

Khu, S.T. and Werner, M.G.F. (2003). Reduction of Monte-Carlo simulation runs for uncertainty estimation in hydrological modelling, *Hydrology and Earth System Sciences*, 7, 5, 680-692.

Kim, Y.-O., Jeong, D., and Ko, I. H. (2006). Combining Rainfall-Runoff Model Outputs for Improving Ensemble Streamflow Prediction, *Journal of Hydrologic Enginering*, 11, 6, 578-588.

Kisi, O. (2008), Stream flow forecasting using neuro-wavelet technique, *Hydrological Processes*, 22, 4142–4152,

Kollat, J. B., Reed, P. M. and Wagener, T. (2012). When are multiobjective calibration trade-offs in hydrologic models meaningful? *Water Resources Research*, 48, 3520.

Krauße, T. and Cullmann, J. (2012). Towards a more representative parametrisation of hydrologic models via synthesizing the strengths of Particle Swarm Optimisation and Robust Parameter Estimation, *Hydrology and Earth System Sciences*, 16, 603-629.

Krauße, T., Cullmann, J., Saile, P., and Schmitz, G H. (2011). Robust multi-objective calibration strategies – chances for improving flood forecasting. Hydrology and Earth System Sciences Discussion, 8, 3693-3741.

Krzysztofowicz, R. (1999). Bayesian theory of probabilistic forecasting via deterministic hydrologic model, *Water Resources Research*, 35, 9), 2739–2750

Krzysztofowicz, R. (2002). Probabilistic flood forecast: Bounds and approximations, *Journal of Hydrology* (Amsterdam), 268(1 -4), 41-55.

Kuczera, G, and Parent, E. (1998). Monte Carlo assessment of parameter uncertainy in conceptual catchment models: the Metropolis algorithm, *Journal of Hydrology*, 211, 69-85.

Kuczera, G. (1997). Efficient subspace probabilistic parameter optimization for catchment models, *Water resources research*, 33, 1, 177-185.

Kuzmin, V., Jun, S. D. and Koren, V. (2008). Fast and efficient optimization of hydrologic model parameters using a priori estimates and stepwise line search, *Journal of Hydrology*, 353, 109-128.

Laloy, E., and Vrugt, J. A. (2012), High-dimensional posterior exploration of hydrologic models using multiple-try DREAM (ZS) and high-performance computing, *Water Resources Research*, 48 , W01526.

Laloy, E., Fasbender, D., and Bielders, C. L. (2010). Parameter optimization and uncertainty analysis for plot-scale continuous modeling of runoff using a formal Bayesian approach, *Journal of Hydrology*, 380, 82-93.

Lee, G., Tachikawa, Y, Takara, K. (2007). Quantification of parameter uncertainty in distributed Rainfall-Runoff Modeling, *Annuals of Disaster Prevention Research Institute*, Kyoto University, 50.

Li, L., Xia, J., Xu, C.-Y., Singh, V.P., (2010). Evaluation of the subjective factors of the GLUE method and comparison with the formal Bayesian method in uncertainty assessment of hydrological models, *Journal of Hydrology*, 390(3-4), 210-221.

Lindström, G., Johansson, B., Persson, M., Gardelin, M., and Bergström, S. (1997). Development and test of the distributed HBV-96 hydrological model, *Journal of Hydrology*, 201, 272-288.

Liu, Y. B., Batelaan, O., De Smedt, F., Poórovó, J., and Velcicka, L. (2005). Automated calibration applied to a

GIS based flood simulation model using PEST. Floods, from defense to management, J. van Alphen, E. van Beek and M. Taal, eds., Taylor and Francis Group, London, 317-326.

Lu, S., Kayastha, N, van Griensven, A., Thodsen, H., Andersen, E. (2012). Multi-objective calibration for comparing channel sediment routing models in SWAT, *Journal of Environmental Quality*, doi:10.2134/jeq2011.0364

Madsen H. (2000). Automatic calibration of a conceptual rainfall–runoff model using multiple objectives, *Journal of Hydrology*, 235, 276-288.

Madsen, H., Wilson, G. and Ammentorp, H. (2002), Comparison of differ-ent automated strategies for calibration of rainfall-runoff models, *Journal of Hydrology*, 261, 48–59.

Maheepala, S., Leighton, B., Mirza, F., Rahilly, M. and Rahman, J. (2005). HydroPlanner - A linked modelling system for water quantity and quality simulation of total water cycle, In Zerger, A. and Argent, R.M. (eds) MODSIM 2005 International Congress on Modelling and Simulation Society of Australia and New Zealand, December 2005, 683-689. ISBN: 0-9758400-2-9.

Maier, H. R., Jain, A., Dandy, G. C. and Sudheer, K. P. (2010). Methods used for the development of neural networks for the prediction of water resource variables in river systems: Current status and future directions, *Environmental Modelling & Software*, 25, 891-909.

Maier, H. M. and Dandy, G. C. (2000). Neural networks for the prediction and forecasting of water resources variables: a review of modelling issues and applications,*Environmental Modelling and Software*, 15, 1, 101-124.

Mantovan, P. and Todini, E. (2006). Hydrological forecasting uncertainty assessment: Incoherence of the GLUE methodology. *Journal of Hydrology*, 330, 1-2, 368-381.

Maringanti, C., Chaubey, I. and Popp, J. (2009). Development of a multiobjective optimization tool for the selection and placement of best management practices for nonpoint source pollution control. *Water Resources Research*, 45.

Marshall, L., Nott, D., and Sharma, A. (2007). Towards dynamic catch-ment modelling: a Bayesian hierarchical mixtures of experts framework, *Hydrological Processes*, 21, 847-861.

Marshall, L., Nott, D., Sharma, A. (2004). A comparative study of Markov chain Monte Carlo methods for conceptual rainfall-runoff modeling, *Water Resources Research*, 40:W02501.

McKay, M. D., Beckman, R. J. and Conover, W. J. (1979). A comparison of three methods for selecting values of input variables in the analysis of output from a computer code, *Technometrics*, 21, 2, 239-245.

McLeod, A. I., Noakes, D. J., Hipel, K. W., Thompstone, R. M. (1987). Combining hydrologic forecast, *Journal of Water Resource. Planning and Management*, 113, 1, 29-41.

McMillan, H. and Brasington, J., (2008). End-to-end flood risk assessment: A coupled model cascade with uncertainty estimation. *Water Resources Research*, 44, 3.

McMillan, H., and Clark, M. (2009). Rainfall-runoff model calibration using informal likelihood measures within a Markov chain Monte Carlo sampling scheme, *Water Resources Research*, 45, W04418.

Melching, C. S. (1995). Reliability estimation. In: V.P. Singh (ed.), *Computer Models of Watershed Hydrology*, Water Resources Publications, Highlands Ranch, CO, USA, 69-118.

Merwade, V., Olivera, F., Arabi, M., and Edleman, S. (2008). Uncertainty in flood inundation mapping — Current issues and future directions, *Journal of Hydrology Engineering*, 13, 7, 608 -620.

Metropolis, N., Rosenbluth, A., Rosenbluth, M, Teller, A., Teller, E. (1953). Equations of state calculations by fast computing machines, *Journal of Chemical Physics*, 21, 1087-1092.

Minasny, B., Vrugt, J. A., and McBratney, A. B. (2011). Confronting uncertainty in model-based geostatistics using Markov Chain Monte Carlo simulation, *Geoderma*, 163,3, 150-162.

Mitchell T. (1997). *Machine Learning*. McGraw-Hill, Singapore, 414.

Montanari A. (2005). Large sample behaviors of the generalized likelihood uncertainty estimation (GLUE) in assessing the uncertainty of rainfall-runoff simulations, *Water Resources Research*, 41, W08406.

Montanari, A. (2011). Uncertainty of Hydrological Predictions. In: Peter Wilderer (ed.) Treatise on Water Science, 2, 459–478, Oxford: Academic Press.

Montanari, M., Hostache, R., Matgen, P., Schumann, G., Pfister, L., and Hoffmann, L.(2009). Calibration and sequential updating of a coupled hydrologic-hydraulic model using remote sensing-derived water stages, *Hydrology and Earth System Science*, 13, 367-380.

Montanari, A., and Koutsoyiannis, D. (2012). A blueprint for process based modeling of uncertain hydrological systems, *Water Resources Research*, 48, 9.

Monteith, J. L. (1965). Evaporation and environment. *In Symposia of the Society for Experimental Biology*, 19, 205-23, 4.

Moore, B. (2002). Special Issue: HYREX: The hydrological radar experiment *Hydrology and Earth System Sciences*, 4(4), 521-522. Nielsen, S. and Hansen, E. (1973). Numerical simulation of the rainfall-runoff process on a daily basis, *Nordic Hydrology*, 4, 3, 171-190.

Mousavi, S. J., Abbaspour, K. C., Kamali, B., Amini, M., and Yang, H. (2012). Uncertainty-based automatic calibration of HEC-HMS model using sequential uncertainty fitting approach, *Journal of Hydroinformatics*, 14, 2, 286-309.

Nash, J., and Sutcliffe, J. (1970). River flow forecasting through conceptual models – Part I – A discussion of principles, *Journal of Hydrology*, 10, 282-290.

Nasr, A. and Bruen, M. (2008). Development of neuro-fuzzy models to account for temporal and spatial variations in a lumped rainfall–runoff model, *Journal of Hydrology*, 349, 277-290.

Nayak, P.C., Sudheer, K.P., Rangan, D.M., Ramasastri, K. S. (2005). Short-term flood forecasting with a neurofuzzy model, *Water Resources Reaserch*, 41, 2517–2530.

Neal, J. C., Fewtrell, T. J. and Trigg, M. A. (2009). Parallelisation of storage cell flood models using OpenMP, *Environmental Modelling Software*, 24, 7, 872-877.

Oudin, L., Andreassian, V., Mathevet, T., Perrin, C. and Michel, C. (2006). Dynamic averaging of rainfall-runoff model simulations from complementary model parameterizations, *Water Resources Research*, 42, 7.

Pappenberger F, Harvey H, Beven K, Hall J, Meadowcroft I. (2006). Decision tree for choosing an uncertainty analysis methodology: a wiki experiment http://www.floodrisknet.org.uk/methods, http://www.floodrisk.net. *Hydrological Processes*, 20, 3793-3798.

Pappenberger, F, Ghelli, A, Buizza, R, Bodis, K. (2009). The skill of probabilistic precipitation forecasts under observational uncertainties within the generalized likelihood uncertainty estimation framework for hydrological applications, *Journal of Hydrometeorology*, 10, 807-819.

Pappenberger, F., Beven, K. J., Hunter, N. M., Bates, P. D., Gouweleeuw, B. T., Thielen, J. and de Roo, A. P .J. (2005a). Cascading model uncertainty from medium range weather forecasts (10 days) through a rainfall–runoff model to flood inundation predictions within the European Flood Forecasting System (EFFS), *Hydrology and Earth System Sciences*, 9, 4, 381-393.

Pappenberger, F., Beven, K., Horritt, M. S. and Blazkova, S., (2005b). Uncertainty in the calibration of effective roughness parameters in HEC-RAS using inundation and downstream level observations, *Journal of Hydrology*, 302,(1-4), 46-69.

Pappenberger, F., Beven, K.J., (2006). Ignorance is bliss: 7 reasons not to use uncertainty analysis. *Water Resources Research*, 42, 5, W05302.

Pappenberger, F., Frodsham, K., Beven, K. J., Romanowicz, R., and Matgen, P. (2007). Fuzzy set approach to calibrating distributed flood inundation models using remote sensing observation, *Hydrology and Earth System Sciences*, 11, 739-752.

Pappenberger, F., Matgen, P., Beven, K., Henry, J., Pfister, L., and Fraipont de, P. (2006). Influence of uncertain boundary conditions and model structure on flood inundation predictions, *Advances in Water Resources*, 29, 10, 1430-1449.

Parrish, M., Moradkhani, H.and DeChant, C. (2012). Towards reduction of model uncertainty: Integration of Bayesian model averaging and data assimilation, *Water Resources. Research.*, 48, W03519,

Pender G., and Neelz S. (2007). Use of computer models of flood inundation to facilitate communication in flood risk management, *Environmental Hazards*, 7, 106-114.

Principe, J., Euliano, N. and Lefebvre, W. (1999). *Neural and adaptive systems: Fundamentals through simulations with CD-ROM.* John Wiley & Sons, New York, NY, USA.

Pushpalathaa, R., Perrina, C., Nicolas Le M., Andréassiana, V. (2012). A review of efficiency criteria suitable for evaluating low-flow simulations, *Journal of Hydrology*, 420, 421, 171-182.

Quinlan, J. R. (1992). Learning with continuous classes, *Proceeding. of The 5th Australian Joint Conference on AI*, World Scientific, Singapore, 343-348.

Rahman, J. M., Seaton, S. P., Perraud, J-M, Hotham, H., Verrelli, D. I. and Coleman, J. R., (2003). It's TIME for a New Environmental Modelling Framework. *Proceedings of MODSIM 2004 International Congress on Modelling and Simulation*, Townsville, Australia, 14-17.

Refsgaard, J. C. (1996). Terminology, modelling protocol and classification of hydrological model codes. In: M.B. Abbott and J.C. Refsgaard (eds.), *Distributed Hydrological Modelling*, Kluwer Academic Publishers, Dordrecht, The Netherlands, 17-39.

Remegio B., Jr., and Whittaker, G. W. (2007). Automatic Calibration of Hydrologic Models with Multi-Objective Evolutionary Algorithm and Pareto Optimization, *Journal of American Water Resources Association*, 43, 4, 981-989.

Romanowicz, R. and Beven, K. (1996). Bayesian calibration of flood inundation models. In: Floodplain Processes (ed. by M. G. Anderson, D. E. Walling & P. D. Bates), 297-318. J. Wiley & Sons Ltd., Chichester, UK

Romanowicz, R.J., Young, P.C., and Beven, K.J. (2006). Data assimilation and adaptive forecasting of water levels in the river Severn catchment, United Kingdom, *Water Resources Research*, 42, 6.

Saint, K., and Murphy, S. (2010). End-to-End Workflows for Coupled Climate and Hydrological Modeling, Proceedings of the International Environmental Modelling and Software Society (iEMSs) 2010 International Congress on *Environmental Modelling and Software Modelling for Environment's Sake*, Fifth Biennial Meeting, Ottawa, Canada.

Schaake, J. C. (1990). From climate to flow, in Climate Change and U.S. Water Resources, edited by P. E.aggoner, chapter. 8, 177– 206, John Wiley, New York.

Schoups, G., and Vrugt, J. (2010). A formal likelihood function for parameter and predictive inference of hydrologic models with correlated, heteroscedastic, and non -Gaussian errors, *Water Resources Research*, 46, W10531

Schumann, G., Henry, J., Hoffmann, B. L., Pfister, ., L. Pappenberger, F., and Matgen, P. (2005). Demonstrating the high potential of remote sensing in hydraulic modelling and flood risk management , paper presented at *Annual Conference of the Remote Sensing and Photogr ammetry Society* With the NERC Earth Observation Conference, Remote Sens. and Photogramm. Soc., Portsmouth, U. K.

Schumann, G., Bates, P. D., Horritt, M. S., Matgen, P., and Pappenberger, F. (2009). Progress in integration of remote sensing–derived flood extent and stage data and hydraulic models, *Reviews of Geophysics*, 47, 4.

See, L. and Openshaw, S. (1999). Applying soft computing approaches to river level forecasting, *Hydrological Sciences Journal*. 44, 5, 763.

See, L. and Openshaw, S. (2000). A hybrid multi-model approach to river level forecasting, *Hydrological Sciences Journal*, 45, 523-536.

Seibert, J. (1997). HBV light version 1.2, User's manual, Uppsala University, Dept. of Earth Science, Hydrology, Uppsala

Shafii, M. and Smedt, F.D. (2009). Multi-objective calibration of a distributed hydrological model (WetSpa) using a genetic algorithm, *Hydrology and Earth System Sciences*, 13, 2137-2149.

Sharma, R. H., and Shakya, N. M. (2006). Hydrological changes and its impact on water resources of Bagmati watershed, Nepal. *Journal of Hydrology*, 327, 3, 315-322.

Shamseldin, A. Y., O'Connor, K. M. and Nasr, A. E. (2007). A comparative study of three neural network forecast combination methods for simulated river flows of different rainfall–runoff models, *Hydrological Sciences Journal*, 52, 896-916.

Shamseldin, A. Y., O'Connor, K. M., and Liang, G. C. (1997). Methods for Combining the Output of Different Rainfall-Runoff Models, *Journal of Hydrology*, 197, 203-229.

Shamseldin, A.Y. and O'Connor, K.M. (1999). A real-time combination Method for the outputs of different rainfall-runoff models. *Hydrological Sciences Journal*, 44, 895-912.

Shannon, C .E. (1948). A mathematical theory of communication. *The Bell System Technical Journal*, 27, 379-423, 623-656.

Shepard, D. (1968, January). A two-dimensional interpolation function for irregularly-spaced data, In *Proceedings of the 1968 23rd ACM national conference*,517-524.

Skahill, B. E., and Doherty, J. (2006). Efficient accommodation of local minima in watershed model calibration, *Journal of Hydrology*, 329, 1, 122-139.

Solomatine, D. P., & Shrestha, D. L. (2004, July). AdaBoost. RT: a boosting algorithm for regression problems. In Neural Networks, 2004. *Proceedings. 2004 IEEE International Joint Conference on* IEEE, 2, 1163-1168.

Singh, V.P. (1995). *Computer models of watershed hydrology*. Water Resources Publication, Highlands Ranch, CO, USA

Shrestha, D. L. and Solomatine, D. P. (2008). Data-driven approaches for estimating uncertainty in rainfall-runoff modelling, *Journal of River Basin Management*, 2, 109-122.

Shrestha, D. L. and Solomatine, D.P. (2006). Machine learning approaches for estimation of prediction interval for the model output, *Neural Networks*, 19, 225-235.

Shrestha, D. L., Kayastha, N., Solomatine, D. P. (2009). A novel approach to parameter uncertainty analysis of hydrological models using neural networks, *Hydrology and Earth System Sciences*, 13, 1235-1248.

Shrestha, D. L., Kayastha, N., Solomatine, D. P. and, Price R. K. (2013) Encapsulation of parametric uncertainty statistics by various predictive machine learning models: MLUE method, *Journal of Hydroinformatics*, 16, 1, 95-113

Shi, X. H., Liang, Y. C., Lee, H. P., Lu, C., and Wang, L. M. (2005). An improved GA and a novel PSO-GA-based hybrid algorithm, *Information Processing Letters*, 93,5, 255-261.

Smith, T. J., and Marshall, L. A. (2008). Bayesian methods in hydrologic modeling: A study of recent advancements in Markov chain Monte Carlo techniques, *Water Resources Research*, 44, W00B05.

Solomatine, D . P. (1999). Two strategies of adaptive cluster covering with descent and their comparison to other algorithms, *Journal of Global Optimization*, 14, 1, 55-78.

Solomatine, D. (2005). Data-driven modeling and computational intelligence methods in hydrology. In: M. Anderson (ed.), *Encyclopedia of Hydrological Sciences*, John Wiley & Sons, New York, NY, USA.

Solomatine, D. P. (2006). Optimal modularization of learning models in forecasting environmental variables, *Proceedings of the iEMSs 3rd Biennial Meeting: "Summit on Environmental Modelling and Software"* (A. Voinov, A. Jakeman, A. Rizzoli, eds.), Burlington, USA, July 2006.

Solomatine, D. P. and Shrestha, D. L. (2006). Experiments with AdaBoost.RT, an Improved Boosting Scheme for Regression 6, *Neural computing*, 18, 7, 1678-1710.

Solomatine, D. P. and Siek, D. P. (2006). Modular learning models in forecasting natural phenomena, *Neural Networks*, 19, 215–224.

Solomatine, D. P. and Dulal, K.N. (2003). Model tree as an alternative to neural network in rainfall-runoff modelling, *Hydrological Sciences Journal*, 48, 3, 399-412.

Solomatine, D. P. and Ostfeld, A. (2008). Data-Driven Modelling: Some Past Experiences and New Approaches, *Journal of Hydroinformatics*, Special Issue on Data Driven Modeling and Evolutionary Optimization for River Basin Management, 10, 1, 3-22.

Solomatine, D. P. and Xue, Y. (2004). M5 model trees compared to neural networks: application to flood forecasting in the upper reach of the Huai River in China, *Journal of Hydraulic Engineering*, 6, 491-501.

175

Solomatine, D. P., Dibike, Y. and Kukuric, N. (1999). Automatic Calibration of Groundwater Models Using Global Optimization Techniques, *Hydrological Science Journal*, 44, 6, 879-894.

Spear, R. and Homberger, G. (1980). Eutrophication in Peel Inlet. II. Identification of critical uncertainties via generalized sensitivity analysis, *Water Resources*. 14, 43-49

Stedinger, J. R., Vogel, R. M., Lee, S. U. and Batchelder, R. (2008). Appraisal of the generalized likelihood uncertainty estimation (GLUE) method, *Water Resources Research*, 44,W00B06.

Sugawara, M. (1967). The flood forecasting by a series storage type model.Int. Symp. Floods and their Computation, IAHS Publ. 85, IAHS, Press, Wallingford, UK. 1-6.

Sugawara, M. (1995). Tank model. Chapter 6. In Computer Models of Watershed Hydrology, Singh, V. P. (Ed.), Water Resources Publications, Littleton, CO.

Tabios, G. and Salas, J. (1985). A comparative analysis of techniques for spatial interpolation of precipitation, *Water Resources Bulletin*, 21, 365-380.

Tang, Y., Reed, P., and Kollat, J. B. (2006). Parallelization Strategies for Rapid and Robust Evolutionary Multiobjective Optimization in Water Resources Applications, *Advances in Water Resources.*, 30, 3, 335-353.

Thielen, J., Bartholmes, J., Ramos, M.-H., de Roo, A. (2009). The European Flood Alert System –Part 1: Concept and development, *Hydrology and Earth System Science*, 13, 125.

Thiemann, M., Trosser, M., Gupta, H. and Sorooshian, S. (2001). Bayesian recursive parameter estimation for hydrologic models, *Water Resources Research*, 37, 10, 2521-2536.

Thiessen, A. H. (1911). Precipitation averages for large areas, *Monthly weather review*, 39, 7, 1082-1089.

Thyer, M., Kuczera, G. and Bates, B. C. (1999). Probabilistic optimization for conceptual rainfall–runoff models: a comparison of the shuffled complex evolution and simulated annealing algorithms. *Water Resources Research*, 35, 3, 767-773.

Todini, E. and Mantovan, P. (2007). Comment on: 'On undermining the science?' by Keith Beven. *Hydrological Processes*, 21, 12, 1633-1638.

Todini, E. and Wallis, J. R. (1977). Using CLS for Daily or Longer Period Rainfall-Runoff Modelling. In Mathematical Models for Surface Water Hydrology, T. A. Ciriani, U. Maione, J. R. Wallis (Eds.), John Wiley & Sons, London, 149-168.

Toth E. (2009). Classification of hydro-meteorological conditions and multiple artificial neural networks for streamflow forecasting, *Hydrology and Earth System Sciences*, 13, 1555-1566.

Uhlenbrook, S., Seibert, J., Leibundgut, C. and Rodhe, A. (1999). Prediction uncertainty of conceptual rainfall-runoff models caused by problems in identifying model parameters and structure, *Hydrological Sciences Journal*, 44, 5, 779-797.

van Griensven, A. and Bauwens, W. (2003). Multi-objective auto-calibration for semi-distributed water quality models, *Water Resources Research*, 39, 10, 1348.

van Griensven, A., Meixner, T., Grunwald, S., Srinivasan, R. (2008). Fit-for-purpose uncertainty versus calibration uncertainty in model-based decision making, *Hydrological Sciences Journal*, 53,5, 1090-1103

van Werkhoven, K., Wagener, T., Reed, P. and Tang, Y. (2009). Senitivity guided reduction of parametric dimensionality for multi-objective calibration of watershed models, *Advances in Water Resources*, 32, 1154-1169.

Vel´azquez, J. A., Anctil, F., and Perrin, C. (2010). Performance and reliability of multimodel hydrological ensemble simulations based on seventeen lumped models and a thousand catchments, *Hydrology and Earth System Sciences*, 14, 2303-2317.

Viney, N., Bormann, H., Breuer, L., Bronstert, A., Croke, B. F. W., Frede, H., Gr¨aff, T., Hubrechts, L., Huisman, J., Jakeman, A., Kite, G., Lanini, J., Leavesley, G., Lettenmaier, D., Lindstr¨om, G., Seibert, J., Sivapalan, M. and Willems, P. (2009). Assessing the impact of land use change on hydrology by ensemble modelling (LUCHEM) II: Ensemble combinations and predictions, *Advances in Water Resources*, 32, 147-158.

Voinov, A., and Cerco, C. (2010). Model integration and the role of data. *Environmental Modelling and Software*, 25, 8, 965-969.

Vrugt, J., Diks, C., Gupta, H., Bouten, W., Verstraten, J. (2005). Improved treatment of uncertainty in hydrologic modeling: Combining the strengths of global optimization and data assimilation. *Water Resources Research*, 41, W01017.

Vrugt, J. and Bouten, W. (2002). Validity of first-order approximations to describe parameter uncertainty in soil hydrologic models, *Soil Science Society of American Journal*, 66 ,6, 1740-1752.

Vrugt, J. and Robinson, B. (2007). Treatment of uncertainty using ensemble methods: Comparison of sequential data assimilation and Bayesian model averaging, *Water Resources Research*, 43, W01411.

Vrugt, J., Diks, C., Gupta, V., Bouten, W., Verstraten, J. (2005). Improved treatment of uncertainty in hydrologic modeling: Combining the strengths of global optimization and data assimilation. *Water Resources Research*, 41, W01017.

Vrugt, J., Gupta, V., Bouten, W., Sorooshian, S., (2003). A shuffled complex evolution metropolis algorithm for optimization and uncertainty assessment of hydrologic model parameters. *Water Resources Research*, 39, 8, 1201.

Vrugt, J., ter Braak C. J. F., Gupta, V. and Robinson, A., (2008). Equifinality of formal (DREAM) and informal (GLUE) Bayesian approaches in hydrologic modeling?, *Stochastic Environmental Research and Risk Assessment*, 23, 7, 1011-1026.

Vrugt, J.A., Braak, C.J.F., Diks, C.G.H., Robinson, B.A., Hyman, J.M., Higdon, D., Modeling, M. (2009). Accelerating Markov Chain Monte Carlo Simulation by Differential Evolution with Self-Adaptive Randomized Subspace Sampling, *Journal of Nonlinear Sciences & Numerical Simulation* ,10, 271-288.

Vrugt, J.A., Schoups, G., Hopmans, J.W., Young, C., Wallender, W.W., Harter, T. and Bouten, W. (2004). Inverse modeling of large scale spatially distributed vadose zone properties using global optimization, *Water Resources Research* ,40. WR002706.

Wang, W., Cheng, C, Chau, K, Xu, D (2012). Calibration of Xinanjiang model parameters using hybrid genetic algorithm based fuzzy optimal model, *Journal of Hydroinformatics*, 14, 3, 784.

Wang, Q.J. (1991). The genetic algorithm and its application to calibrating conceptual rainfall-runoff models, *Water Resources Research*, 27 ,9, 2467-2471.

Warmink, J., Van der Klis, H., Booij, M., and Hulscher, S. (2011). Identification and quantification ofuncertainties in a hydrodynamic river model using expert opinions. *Journal of Water Resources Management*, 25, 2, 601-622.

Westerberg, I. K., Guerrero, J. L., Younger, P. M., Beven, K. J., Seibert, J., Halldin, S., Freer, J. E., and Xu, C. Y. (2011). Calibration of hydrological models using flow-duration curves, *Hydrology and Earth System Sciences*, 15, 2205-2227.

Willems, P. (2009). A time series tool to support the multi-criteria performance evaluation of rainfall-runoff models, *Environmental Modelling and Software*, 24, 311-321.

Witten, I. H. and Frank, E. (2000). Data Mining: Practical Machine Learning Tools and Techniques with Java implementations, 239, Morgan Kaufmaan, San Francisco, CA, USA

WMO (1975). *Intercomparison of conceptual hydrological models used in operational hydrological forecasting. Geneva.* Operational Hydrology Report No. 7, WMO No. 429, World Meteorological Organization, Geneva, Switzerland.

WMO (1986). Intercomparison of Models of Snowmelt Runoff. Operational Hydrology Report No. 23, WMO Report No. 646, *World Meteorological Organization*, Geneva, Switzerland,

Xiong, L. and O'Connor, K. (2008). An empirical method to improve the prediction limits of the GLUE methodology in rainfall-runoff modeling, *Journal of Hydrology*, 349, 115-124.

Xiong, L., Shamseldin, A. Y. and O'Connor, K. M. (2001). A non-linear combination of the forecasts of rainfall-runoff models by the first-order Takagi-Sugeno fuzzy system, *Journal of Hydrology*, 245, 1-4), 196-217.

Xiong, L., Wan, M., Wei, X., O'connor, M. (2009). Indices for assessing the prediction bounds of hydrological models and application by generalised likelihood uncertainty estimation, *Hydrological Sciences Journal*, 54, 5.

Xu, C. Y. (1999). Climate change and hydrologic models: A review of existing gaps and recent research developments, *Water Resources Management*, 13, 369-382.

Yang, J., Reichert, P., Abbaspour, K.C., Xua, J., Yang, H. (2008). Comparing uncertainty analysis techniques for a SWAT application to the Chaohe basin in China, *Journal of Hydrology*, 358, 1-23.

Yapo P., Gupta H., Sorooshian S. (1996). Automatic calibration of conceptual rainfall-runoff models: sensitivity to calibration data. *Journal of Hydrology*, 181, 23-48.

Yapo P., O., Gupta H., V., and Sorooshian, S. (1998). Multi-objective global optimization for hydrologic models, *Journal of Hydrology*, 204, 1-4, 30, 83-97.

Yatheendradas, S., Wagener, T., Gupta, H., Unkrich, C., Goodrich, D., Schaffner, M., and Stewart, A. (2008). Understanding uncertainty in distributed flash-flood forecasting for semi-arid regions, *Water Resources Research*, 44, 5.

Young, P. and Ratto, M. (2009). A unified approach to environmental systems modeling. *Stochastic Environmental Research and Risk Assessment*, doi: 10.1007/s00477-008-0271-1.

Zhang X. and Zhao, K. (2012). Bayesian Neural Networks for Uncertainty Analysis of Hydrologic Modeling and: A Comparison of Two Schemes, *Water Resource Management*, 26, 2365-2382.

Zhang, X., Srinivasan, R. and Bosch, D. (2009). Calibration and uncertainty analysis of the SWAT model using genetic algorithms and Bayesian model, averaging, *Journal of. Hydrology*, 374, 3-4, 307-317.

Zhang, X., Srinivasan, R., Zhao, K. and van Liew , M. (2008). Evaluation of global optimization algorithms for parameter calibration of a computationally intensive hydrologic model, *Hydrological processes*, 23, 3, 430-441.

Zhao, R. J. (1977). Flood forecasting method for humid regions of China, *East China College of Hydraulic Engineering, Nanjing*, China.

LIST OF ACRONYMS

ACCO	Adaptive Cluster Covering Algorithm
ALI	Advanced Land Imager
AM	Adaptive Metropolis-Hastings
AMALGAM	A MultiALgorithm Genetically Adaptive Multiobjective
AMI	Average Mutual Information
ANNs	Artificial Neural Networks
AROPE	Advanced Robust Parameter Estimation
AWBM	Australian water balance model
BA	Bayesian approach
BMA	Bayesian model averaging
BNN	Bayesian neural networks
CDF	Cumulative density function
CGA	Chaos genetic algorithm
CGASA	Chaos genetic algorithm simulated annealing.
CLS-Ts	Constrained Linear Systems with a single Threshold
CM	Committee model
CoC	Correlation coefficient average mutual information
CR	Contain ratio
CRPS	Continuous rank probability score
CSA	Clustering and Simulated Annealing
CV	Cross Validation
DDM	Data Driven Models
DEM	Digital Elevation Model
DEMC	Differential Evolution Markov Chain
DRAM	Delay rejection Metropolis-Hastings
DREAM	Delay rejection evolution adaptive Metropolis
FC	Fuzzy committee
FL	Fuzzy logic
GA	Genetic Algorithm
GLUE	Generalized Likelihood Uncertainty Estimation
HBV	Hydrologiska Byråns Vattenbalansavdelning
HF	high flows model
HSPF	Hydrological Simulation Program-FORTRAN model
IBL	Instance based learning
KBES	Knowledge-based expert system;
LF	Low flow model
LHS	Latin Hypercube Sampling
LPM	Linear Perturbation Model
LREB	Little River Experimental Basin;
LREW	Little River Experimental Watershed;
LTF	Linear Transfer Function () and
LTF	Linear Transfer Function
SOM	Self Organizing Maps
MLP	multi-layer perceptron neural
BA	Bayesian approach
FL	Fuzzy logic
MMSE	Modified Multimodel Superensemble
BMA	Bayesian model averaging
SWB	Simple Water Balance model
SNN	Simple Neural Network
RBFNN	Radial Basis Function Neural Network
MLPNN	Multi-Layer Perceptron Neural Network
LVGFM	Linearly Varying Variable Gain Factor Model
LWR	Locally weighted regression
M3SE	Modified Multimodel Superensemble

MBF	Outputs-based committee models,
MBI	Inputs-based committee models
MC	Monte Carlo
MCEW	Mahantango Creek Experimental Watershed
MCMC	Markov Chain Monte Carlo
MFM	Fuzzy committee model
MG -	Meta-Gaussian
mGLUE	modified Generalized likelihood uncer-tainty estimation
MH	Metropolis-Hastings algorithm
MHBC	Metropolis-Hastings algorithm of Bates and Campbell;
MHBU	Metropolis-Hastings algorithm with block update
MHSS	Metropolis-Hastings algorithm with single-site update
MLANN	Multi-Layer Perceptron Neural Network
MLUE	Machine Learning Uncertainty Estimation
MM	Modular Model
MMSE	Multimodal Superensemble
MP	Multistart Powell
MQN	(Mq-N) Multistart Quasi-Newton
MS	Multistart simplex
MS2	Multiple Starting Points 2
mSFB	modified SFB Model
MSSE-PSO	Master–Slave Swarms Shuffling Evolution Algorithm PSO
MSV	States based committee model
MT	Model Tree
NERC	Natural Environment Research Council
NNs	Neural Networks
NSGA II	Non-dominated Sorted Genetic Algorithms II
NSGAX	multi-objective optimization
pdf	Probability density function
PEST	Parameter estimation package
PSO	Particle Swarm Optimization
PSSE-PSO	Parallel Swarms Shuffling Evolution Algorithm PSO
RBFNN	Radial Basis Function Neural Network
RCEW	Reynolds Creek Experimental Watershed
ROPE	Robust parameter estimation
SAM	Simple average method
SA-SX	Simulated-Annealing Algorithm
SCE	Shuffled Complex Evolution
SCEMUA	Shuffled Complex Evolution Metropolis
SLM	Simple Linear Model
SM	Single hydrological model (Qs)
SMA	Sacramento Soil Moisture Accounting
SMAR	Soil moisture and accounting routing
SNN	Simple neural network
SOBEK	1d-2d Hydrodynamic Model
SODA	Simultaneous Optimization and Data Assimilation
SRS	Scale reduction score
SWAT	Soil Water Assessment Tool
SWB	Simple Water Balance model
TCEF	Tenderfoot Creek Experimental Forest
WAM	weighted average method
WMO	World Metrological organization
YRHB	Yellow River Headwater Basin
YRHW	Yellow River headwaters watershed

LIST OF TABLES

LIST OF FIGURES

SAMENVATTING

Een hydrologisch model is een abstractie van complexe en niet-lineaire fysische processen met als doel om de niet-stationaire waterbeweging in stroomgebieden te kunnen voorspellen. Het succes van dergelijke voorspellingen hangt af van de gekozen modelstructuur, de gekozen parameters, en de kwaliteit van de gegevens die zijn gebruikt. Bij voorspellende (conceptuele hydrologische) modellen wordt over het algemeen aangenomen dat de gebruikte meetgegevens correct zijn, zodat modelvoorspellingen veelal worden gepresenteerd ten opzichte van de meetgegevens waarbij kennis van het onderliggende proces wordt gebruikt om de optimale parameter instellingen te bepalen (kalibratie). Bij modelvoorspellingen moet echter ook rekening worden gehouden met onzekerheid omdat kalibratie en onzekerheid aan elkaar gerelateerd zijn. De betrouwbaarheid van modelresultaten kan niet worden bepaald zonder de onzekerheid te bepalen waarmee de hydrologische respons kan worden voorspeld.

Vaak is het niet mogelijk voor een enkel hydrologisch model om alle hydrologische processen even goed te beschrijven, gelet op de vele verschillende processen die in verschillende mate kunnen plaats hebben. Een 'meervoudige model benadering' opent de mogelijkheid om deze beperkingen op te heffen en de voorspellende waarde van modellen te verbeteren. Een van de mogelijkheden betreft een zogenaamd "comité model" dat in dit proefschrift nader wordt onderzocht. Daarbij worden afzonderlijke modellen voor specifieke hydrologische processen in eenzelfde modelstruktuur samengevoegd tot een nieuw model dat optimaal gebruik maakt van de sterke eigenschappen die elkaars zwakke kanten compenseren.

Speciale aandacht wordt gegeven aan de zogenaamde 'fuzzy comite" aanpak van de hydrologisch modelleren (Solomatine, 2006, Fenicia et al. (2007). In deze aanpak eerst verschillende (stroomgebied) processen worden gekalibreerd teneinde een bepaald proces het best te beschrijven, en ze vervolgens samen te voegen door ze een bepaald (fuzzy) gewicht toe te kennen. Een dergelijk aanpak wordt beschreven door Kayastha et al. (2013) die verschillende typen gewichtsfactoren onderzocht om gespecialiseerde modellen te kunnen kalibreren, alsmede verschillende klassen 'membership functions' om deze modellen te kunnen combineren. De modellen zijn in eerste instantie opgezet om afzonderlijke stromingsaspecten te beschrijven, en vervolgens zodanig samengevoegd dat een betere en nauwkeuriger voorspelling wordt verkregen. In dit proefschrift worden dergelijke modellen aangeduid als "committee models". De gewichten die aan de output van de verschillende gespecialiseerde modellen worden toegekend zijn gebaseerd op de geoptimaliseerde 'fuzzy membership functions'. Het proces wordt in dit proefschrift uitgewerkt. Alle 'committee modellen' vertoonden een beter resultaat dan de enkele (optimale) hydrologische modellen in toepassingen voor het stroomgebied van de Alzette, Bagmati, Brue, en Leaf river. Het bepalen van de gewichten voor een comité-model kan worden gebaseerd op ondermeer de toestandsvariabelen (bodemvochtigheid, basis stroming, etc.), invoer grootheden (neerslag en verdamping) en uitvoerresultaten (gesimuleerde afvoeren) zoals hier beschreven, waarbij is gebleken dat deze gewichtsfactoren in de tijd kunnen variëren afhankelijk van de grootte van de stroming.

De modellen die speciaal waren ontwikkeld voor de trage stromingen in het stroomgebied hadden een relatief grotere fout dan de modellen voor snelle stromingen. Een van de mogelijkheden om het comité-model te verbeteren is om een hybride vorm te ontwikkelen. In dit proefschrift is een comité-model samengesteld uit twee gespecialiseerde modellen (een conceptueel model voor de snelle stromingen en een data-gedreven kunstmatig neuraal netwerk model voor de trage stromingen) waarbij een toegesneden combinatie methodiek (fuzzy membership function) is ontwikkeld. Het concept van hybride comité-modellen is uitgetest op de Bagmati and Leaf stroomgebieden en de resultaten blijken het best van alle comité-modellen.

Onzekerheidsanalyse kan worden gebruikt om de betrouwbaarheid en geloofwaardigheid van hydrologische modelvoorspellingen te verbeteren. Methoden gebaseerd op steekproeven worden veel gebruikt bij hydrologische modellen. Monte Carlo (MC) simulatie is een van de meest populaire steekproeftechnieken waarbij de uitvoer wordt gerelateerd aan de invoer en/of de parameterinstellingen op basis waarvan een kwantitatieve betrouwbaarheidsmaat wordt bepaald, die overigens sterk afhangt van de grootte van de steekproef. De Monte Carlo methode vereist een groot aantal modelsimulaties. Vandaar dat veel aandacht uitgaat naar het ontwikkelen van meer economische steekproeftechnieken waardoor ook met rekenintensieve modellen gewerkt kan worden (bijv. LHS, GLUE, MCMC, enz.). De theorie van Bayes kan worden gebruikt om de a-posteriori verdeling van parameters te berekenen op basis van 'generalized likelihood functions' die verschillende gewichtsfactoren toekennen aan verschillende parameter combinaties of modellen. Een dergelijke aanpak zorgt voor een statistisch betrouwbare beschrijving van de waarschijnlijkheidsverdeling.

In dit proefschrift worden de resultaten gepresenteerd van de onderzochte steekproeftechnieken (MCS, GLUE, MCMC, SCEMUA, DREAM, PSO, and ACCO) om de onzekerheid te bepalen van hydrologische modellen en hun resultaten te vergelijken op basis van onzekerheidsanalyse. Daarbij bleek dat de onzekerheidsanalyse sterk afhangt van de gebruikte steekproeftechniek.

Het schatten van onzekerheid op basis van MC simulaties is een valide benadering voor meetgegevens uit het verleden, maar geldt niet noodzakelijkerwijs voor modelvoorspellingen in de toekomst. Om die te verkrijgen zou het dienstig zijn een economische manier te vinden om de onzekerheid te schatten van de hydrologische systemen. In dit proefschrift worden zelf-lerende computertechnieken (data-gedreven modelleren) gebruikt om de nauwkeurigheid te verbeteren van hydrologische voorspelsystemen. Deze technieken geven echter geen inzicht in de verdelingsfunctie van de onzekerheid. Shrestha et al. (2009, 2013) gaven een aanzet om de verdelingsfunctie te schatten voor hydrologische modellen door karakteristieke invoergroootheden (neerslag-afvoerrelaties uit het verleden) als input te gebruiken voor zelflerende computer technieken om de bijbehorende onzekerheid in de uitvoergroootheden te bepalen. Een korte beschrijving van deze methode om de onzekerheid te bepalen is te vinden in dit proefschrift. De methodiek is getest voor de stroomgebieden van de Bagmati en Brue om de onzekerheidsbanden van de deterministische uitvoer van het conceptuele hydrologische HBV model te schatten. De resultaten laten zien dat dit een efficiënte methode is om onzekerheid te bepalen die bovendien nauwkeurig is. Deze methode is verder onderzocht aan de hand van verschillende steekproeftechnieken voor het bepalen van de onzekerheid in de uitvoer van hydrologische modellen.

De resultaten van verschillende onzekerheidsvoorspellingen (verkregen met behulp van zelflerende computermodellen) kunnen verschillen vanwege: a) het gebruik van verschillende steekproeftechnieken voor verschillende gegevensbestanden; en b) het gebruik van verschillende invoerbestanden om de onzekerheid te bepalen. In dit proefschrift wordt een combinatie van modellen voorgesteld (die een comité vormen) die is toegepast om de onzekerheid te schatten in de afvoeren van een conceptueel model in de stroomgebieden van de Bagmati en Nzoia.

Een belangrijk toepassingsgebied van hydrologische modellen betreft het modelleren van overstromingen. Kennis omtrent de mate van onzekerheid in de voorspellingen is belangrijk voor het nemen van beslissingen om de gevolgen van overstromingen te beperken. Daarbij is het zaak om alle informatie over rivierafvoeren te kennen (afvoerwaarden, ruwheid coëfficiënten, dwarsdoorsneden, waterdiepten) maar ook de ligging van het gebied (topografie) en methoden om overstromingsgebieden goed in kaart te brengen. De afvoer is een belangrijke factor en daarom is kennis van het stromingsgedrag noodzakelijk voor het bepalen van de omvang van een overstroming. Een ingewikkeld overstromingsproces kan worden weergegeven door een reeks (cascade) van afzonderlijke (hydrologische en hydraulische) modellen gekoppeld aan een ruimtelijk landschapsmodel. Dit is echter gemakkelijker gezegd dan gedaan, want dit vereist beschikbaarheid van gegevens, rekenkracht en kennis van de interacties tussen de modellen. Daarbij dient rekening te worden gehouden met de verschillende bronnen van onzekerheid die leiden tot onzekere modelvoorspellingen. Een veelgebruikte techniek is de MC simulatie, die wordt gebruikt om een ensemble van deterministische modelsimulaties te genereren om daar vervolgens een waarde aan toe te kennen gebaseerd op overstromingswaarnemingen in het gebied. Daarbij worden veelal Remote Sensing waarnemingen gebruikt om op een deterministische manier modellen te kalibreren aan de hand van een enkele gebeurtenis. Beter inzicht in de onzekerheid zou kunnen worden verkregen door meerdere overstromingen te simuleren, mits de rekenkracht daarvoor toelatend is. In dit proefschrift zijn het hydrologisch model SWAT en het hydrodynamisch model SOBEK geïntegreerd in een cascade-model om de onzekerheid in het overstromingsgebied te kwantificeren voor het stroomgebied van de Nzoia in Kenya. De modellen zijn ingebed in een high-performance (parallel computing) raamwerk en de uitvoer is gebruikt om de onzekerheid in de overstromingsberekeningen te bepalen en weer te geven in de vorm van een relatieve betrouwbaarheidsband.

Kort samengevat beschrijft dit proefschrift de onderzoeksresultaten van: (i) comité-modellen voor hydrologisch modelleren; (ii) hybride comité-modellen; (iii) het effect van steekproef- technieken op de voorspelling van onzekerheid in hydrologische modellen; (iv) het voorspellen van onzekerheid met behulp van zelflerende (computer) technieken; (v) het gebruik van comite-modellen om onzekerheid te voorspellen; en (vi) het schatten van onzekerheid in overstromingsmodellen. Dit onderzoek draagt bij aan het vakgebied van de hydroinformatica dat zich kenmerkt door verschillende wetenschappelijke disciplines bijeen te brengen: hydrologisch modelleren, hydrodynamisch modelleren, data-gedreven modelleren, het combineren van modellen, het gebruik van hybride modellen, onzekerheidsanalyse en high-performance computing. De resultaten laten zien dat deze methoden bruikbaar zijn in de praktijk. De ontwikkelde software is publiek beschikbaar voor verdere ontwikkeling in onderzoek en toepassing.

Nagendra Kayastha
Delft, Nederland

ACKNOWLEDGEMENT

The most valuable contributor to this research is my promoter/supervisor Prof. Dimitri Solomatine. Dimitri: I sincerely thank you for providing me freedom, and your guidance and encouragement. You have given me not only innovative ideas, critical thought and insightful feedback but also unreserved and decisive support for timely and successful completion of this thesis. I would have never attained this stage without your contributions.

I would also like to express my sincere gratitude to my co-mentor Prof. Ann van Griensven for giving me the opportunity to start this PhD and also for her enthusiastic, industrious supervision and scientific guidance.

My greatest gratitude goes to Dr. Schalk-jan van Andel, Dr. Andrea Jonoski, Dr. Ioana Popescu, Dr. Biswa Bhattacharya, Dr. Leonardo Alfonso Segura and Dr. Gerald Corzo Perez for providing me with scientific discussion and partial funding for this research.

I thank my PhD colleagues—Anuar Ali, Micah Mukolwe, Isnaeni Hartanto, Kun Yan, Juan Carlos Chacon Hurtado, Mario Castro, Maurizio Mazzoleni, Oscar Marquez Calvo, Quan Pan, Zahrah Musa, Oscar Marquez Calvo, Blagoi Delipetrev and other MSc students for entertaining discussions, helpful tips, lots of fun and pleasant conversations.

I will forever be thankful to my former research advisor, Dr. Durga Lal Shrestha. Durga Lal has been helpful in providing frequent advice over the course of my research career.

I would also like express my sincere appreciations to Dr. Shredhar Maskey, Dr. Febrizio Fenicia, Dr Yunqing Xuan, Dr. Michael Siek, Dr. Arlex Sanschez Torres, Dr. Adrian Almoradie, Dr. Girma Ibrahim, Dr. Xuan Zhu, Dr. Sanjay Giri, Dr. Saroj Sharma, Dr. Krishna Khatri, and Ms Manisha Shakya for scientific discussions. We worked together with pleasure. I am also grateful to Jolanda Boots, Maria Laura Sorrentino, Marielle van Erven, Sylvia van Opdorp-Stijlen, Jos Bult, Guy Beaujot, Ewout Heeringa, Anique Karsten and Peter Stroo for assisting in the administrative and practical matters.

Lastly, but not least, my sincere gratitude goes to my family, close relatives, and friends, who gave me their supports and encouragement, particularly to my mother Akhandeswori, brother Pradeep, sister-in-law Subhadra, sister Suryalaxmi, nephew Peter and parents-in-law Lamboder and Uma Raithore. Finally, my special thanks goes to my wife Ushna and sons Ujen and Naman for their unconditional and unlimited support. Without their encouragement, inspiration and support, this journey would not have been possible.

Nagendra Kayastha
October 2014

ABOUT THE AUTHOR

Nagendra Kayastha graduated in Civil Engineering (MSc) from the St. Petersburg State University of Means of Communication, St. Petersburg, Russia, in 1997, with a specialization in Bridge and Tunnel Engineering. He was engaged more than three years as a consulting engineer in Morrison Knudson International Inc.(USA), for the Kaligandaki 'A' Hydropower Project in western Nepal. After completion of this project, he continued working at a consulting company in Nepal and was assigned to various national and international projects, where he gained knowledge in the planning and design of projects, including the water-related projects. In 2005, he joined the MSc degree programme in Water Science and Engineering, specializing in Hydroinformatics at UNESCO-IHE Institute for Water Education, Delft, The Netherlands. His MSc research topic was on the "Novel approaches to uncertainty analysis of hydrological model" which covered new methods for uncertainty prediction of hydrological models using machine learning techniques. After completion of his study, he joined the special research programme at Hydroinformatics and Knowledge Management Department. He was involved in projects of the Delft Cluster Research Programme and in the EU project FLOODsite. He joined the PhD programme of the UNESCO-IHE and the Delft University of Technology, under the supervision of Professor Dimitri Solomatine with the co-supervision of Professor Ann van Griensven (Vrij Universiteit, Brussels) in 2010. He contributed to various projects, namely EnviroGRIDs, WeSenseIT, and MyWater and has assisted in the Master programme in Hydroinformatics. He published more than 15 technical papers in international journals and conferences.

Publications

Kayastha, N., Ye,J., Fenicia,F., Kuzmin,V., and Solomatine, D. P. (2013). Fuzzy committees of specialized rainfall-runoff models: further enhancements and tests, *Hydrology and Earth System Sciences*, 17, 4441-4451.

Shrestha, D. L., Kayastha, N., Solomatine, D. P. and, Price R. K. (2013) Encapsulation of parametric uncertainty statistics by various predictive machine learning models: MLUE method, *Journal of Hydroinformatics*, 16, 1, 95-113

Kayastha, N., van Griensven A, Shrestha, D. L. and Solomatine, D. P. Assessment of numerous methods of parameter uncertainty estimation in hydrological modeling. (in preparation 2014)

Lu. S, Kayastha, N, van Griensven A, Estrup A. (2012). Multi-objective calibration for comparing channel sediment routing models in SWAT, *Journal of Environmental.Quality*, 43,1.

Shrestha, D. L., Kayastha, N. and Solomatine, D. P. (2009). A novel approach to parameter uncertainty analysis of hydrological models using neural networks. *Hydrology and Earth System Sciences*, 13, 1235–1248, 2009.

Kayastha, N and Solomatine, D. P. (2014). Optimal committees of specialised conceptual hydrological models: a comparison of latest methods (in preparation for *Water Resources Research*)

Proceedings of International conferences:

Kayastha, N., Solomatine, D. P, Shrestha, D. L., (2014). Prediction of hydrological models' uncertainty by a committee of machine learning-models, *11ᵗʰ International conference on Hydroinformatics*, New York USA

Kayastha, N., and Solomatine, D. P (2014). Committees of specialized conceptual hydrological models: comparative study, 11th International conference on Hydroinformatics, New York USA

Solomatine, D. P., Kayastha, N., Kuzmin, V. A. (2013). Optimization of specialized hydrological models committees adapting to changing regimes, session - Testing simulation and forecasting models in non-stationary conditions, IAHS Symposia, Knowledge for the future, IAHS -IAPSO -IASPEI Joint Assembly, Gothenburg, Sweden

Kayastha, N. van Grienven, A., Yunqing, X., Solomatine, D. P. (2012). Identification of uncertainties in climate change impact on streamflows in the Nzoia catchment, Kenya, 10th International conference on Hydroinformatics, Germany

Solomatine, D. P, Kayastha, N., Ye, J. (2012). Optimal combination of specialized hydrological Models: further enhancements, 10th International conference on Hydroinformatics, Germany (submitted)

Solomatine, D. P., Shrestha, D. L., Kayastha, N., Di Baldassarre, G. (2012). Application of methods predicting model uncertainty in flood forecasting. Second EU conference on FLOODRisk, Rotterdam, The Netherlands.

Kayastha, N., Shenlang, L., Betrie G., Zakayo Z., van Griensven A., Solomatine, D. P. (2011). Dynamic linking of the watershed model SWAT to the multi-objective optimization tool NSGAX, 8th IWA Symposium on Systems Analysis and Integrated Assessment, Watermatex 2011 Spain.

Shenglan, L., Kayastha, N., van Griensven, A., Estrup, A. (2011). Multi-objective calibration of flow and sediment on a small Danish Catchment. International SWAT, Conference & Workshops 2011- Toledo, Spain

Kayastha, N., van Griensven, A., Solomatine, D. P. (2011). Dealing with uncertainties in remotely linked models, OpenWater symposium and workshops, UNESCO-IHE, The Netherlands

Kayastha, N., Shrestha, D. L., Solomatine, D. P. (2010). Experiments with several methods of parameter uncertainty estimation in hydrological modeling, 9th International conference on Hydroinformatics, China

Shrestha, D. L., Kayastha, N., Solomatine, D. P. (2009). Encapsulation of Monte-Carlo uncertainty analysis results in a predictive machine learning model, 8th International Conference on Hydroinformatics, Chile.

Shrestha, D. L., Kayastha, N., Solomatine, D. P. (2009). ANNs and Other Machine Learning Techniques in Modelling Models' Uncertainty: Application to Hydrological Model, 19th International Conference on Artificial Neural Networks, 387-396. Limassol, Cyprus.

Shrestha, D. L., Kayastha, N., Solomatine, D. P. (2009). Parametric uncertainty estimation of a hydrological model using piece-wise linear regression surrogates, 33rd IAHR Congress, Vancouver Canada

Conference abstracts:

Kayastha, N. and Solomatine, D. P. (2014). Committee of machine learning predictors of hydrological models uncertainty, Geophysical Research Abstract, European Geosciences Union. Vol. 15, EGU 2014-7026.

Kayastha, N. and Solomatine, D. P. (2013). Combinations of specialized conceptual and neural network rainfall-runoff models: comparison of performance, *Geophysical Research Abstract, European Geosciences Union*. Vol. 15, EGU 2013-9022.

Kayastha, N., Solomatine, D. P., Shrestha, D. L., van Griensven, A. (2013). Use of different sampling schemes in machine learning-based prediction of hydrological models' uncertainty, *Geophysical Research Abstract, European Geosciences Union*. Vol. 15, EGU 2013-9466.

Ye, J., Kayastha, N., van Andel, S.J., Fenicia, F., Solomatine, D.P. (2012). Experiments with models committees for flow forecasting, *Geophysical Research Abstracts, European Geosciences Union*. Vol. 14, EGU 2012.

Kayastha, N., Shrestha, D.L., Solomatine, D. P. (2011). Influence of sampling strategies on estimation of hydrological models uncertainty, *Geophysical Research Abstracts, European Geosciences Union*. Vol. 13, EGU 2011, 3781.

Kayastha, N., Shrestha, D. L., Solomatine, D. P. (2010).Optimal combinations of specialized conceptual hydrological models. *Geophysical Research Abstract, European Geosciences Union*. Vol. 12, EGU 2010-5345.

Shrestha D. L., Kayastha, N., Solomatine, D. P. (2010). Experiments with encapsulation of Monte Carlo simulation results in machine learning models, *Proceeding Geophysical Research Abstract, European Geosciences Union*. Vol. 12, EGU 2010-12288-1.

Shrestha, D. L., Kayastha, N., Solomatine D. P. (2009). A novel approach to parameter uncertainty analysis of hydrological models: Application of machine learning techniques, *Geophysical Research Abstracts, European Geosciences Union*. Vol. 11, EGU 2009-4676.

Shrestha, D. L., Kayastha, N., Solomatine, D. P. (2009). A novel approach to Monte Carlo-based uncertainty analysis of hydrological models using artificial neural networks, *Geophysical Research Abstracts, European Geosciences Union*. Vol. 11, EGU 2009-4335.

T - #1018 - 101024 - C212 - 240/170/12 - PB - 9781138027466 - Gloss Lamination